Our Commo

Evidence from Genetics, Linguistics,
Archaeology, Genesis, & Pre-Egyptian History
and
How Judeo-Christian Mythology Tried to Erase It

Larry West

*To—
Mom, with
love & gratitude.
Larry*

VANTAGE PRESS
New York

FIRST EDITION

All rights reserved, including the right of
reproduction in whole or in part in any form.

Copyright © 1999 by Larry West

Published by Vantage Press, Inc.
516 West 34th Street, New York, New York 10001

Manufactured in the United States of America
ISBN: 0-533-12815-3

Library of Congress Catalog Card No.: 98-90485

0 9 8 7 6 5 4 3 2 1

**To
Rachel**

Special thanks to Louto.

Contents

List of Figures vii
Preface ix
Prelude xi

1. Genes 1
2. Migrations 3
3. Languages 7
4. Evolution 9
5. Adam and Eve 13
6. Noah's Ark 15
7. The Ice Age and Early Man 16
8. Genesis versus Genetics 19
9. Pre-Egyptian History 21
10. The Early Hebrews and the Cushites 30
11. African Influence on Judeo-Christian Mythology 32
12. When and Where Did the Aryans/Semites Change the Cushite/Egyptian Mythology into Judeo-Christian Mythology and Make It Sound Like Their Own? 38
13. Summary 48

Postlude 51
Author's Apologia 53
References 57
Index 61
Vita 65

List of Figures

Religions Family Tree	front cover
1. Human Genetic Family Tree	xii
2. Earliest Human Migrations	4
3. Correlation of Genetic & Language Family Trees	6
4. Primate Evolutionary Family Tree	10
5. Origins of Organic Evolution	12
6. Table of Nations, Family of Noah	14
7. Maximum Extent of Glaciation	17
8. Ancient Cushite Zodiac	29
9. Hebrew/Egyptian Names and Their Literal Translation	34
10. The Egyptian Admonitions of Innocence	35
11. The Egyptian "Christ Figure"	36
12. Early Caucasian Migrations	37
13. Historical Timelines	39
14. Zodiac of Dendara	66

Preface

The science of genetics has evidence that we modern humans appeared "suddenly" in Africa about 200,000 years ago with no genetic connection to older creatures. Studies of fossils, normally used to promote the Theory of Evolution, also support the same "Out of Africa, Noah's Ark Theory." Sophisticated tools have recently been found deep within Africa that date back 100,000 years. The Book of Genesis and pre-Egyptian history show that this Hamitic (African) predominance carried throughout ancient history, from the kingdoms of Cush and Egypt to the lands of Canaan and Babylon. Cushitic influence has been found from Africa through the Middle East all the way to India, not to mention the Greeks' heavy absorption of Egyptian religion, science, and culture. The Phoenicians, remnants of the original Canaanites, are the acknowledged source of our Western phonetic alphabet.

Books about history and religion often try to prove the religions are historically correct or try to debunk them. This tries to do neither. The author is interested in finding the common ground between history and religion, between knowledge and faith, so that the two don't seem so far apart. In so doing, however, we must distinguish myth from mythology. The strong Egyptian content in Judeo-Christian-Islam mythologies is as obvious as is the attempt by the early Hebrew story-tellers to erase that fact. Why they did so is a historical and cultural issue that we all have had to live with ever since. What at the time was a local struggle for territory between the invading Hebrews and the native black Hamites was documented by the invaders from their point of view. The resulting documents became the foundation for three religions, allegedly inspired by God, but replete with negative portrayals of their African adversaries, thus creating a racial sterotype that has clouded all of Western culture since, and by its influence, much of the rest of the world. Let's look into our past and see what our ancestors were up to.

This collection of material cannot be found in any one place, let alone the same part of town. Most of this information is closeted within each supporting cultural group, each expounding its own views of the world. History tends to be rationalized by exclusion, not inclusion. Whites write about white history, blacks about black history, Christians about Christian history, etc., and so on.

Two accidental exposures got me going—the first, an introduction to the culture of Black American Hebrews, and, the second, a few months later, an article in *Scientific American* (ref. 2) on "Genes, Peoples, and Languages." The similarity between the two struck me like a shot. All of a sudden, the Old Testament and the Torah started to make sense, but it took several years of studies and several painful revelations to reach the last chapter. The most painful was to walk with Moses out of Egypt, home for his people for 250-plus years, and watch him deny his African education, culture, and religion and profess that only his God gave it to him. That set the tone for Western attitudes about Africans and African contributions to history, culture, and religion ever since. Vignettes about Noah and Canaan, Abraham and Hagar, Moab and Israel, Sodom and Gomorrah, and Jezebel and Elijah, to name a few, only amplify the not-so-subliminal signals from our ancient past. Once these ideas were embedded in our religions, they crept into our values and languages, thereby affecting every aspect of Western thought and actions.

Whereas the triumphs of the Judeo-Christian legacy are manifest, the tragedies have not been spoken of with any candor or honesty, save for the revolt of Mohammed . . . until now.

The African influence on our Western culture and religions is awesome, both the positive contributions of the ancient Africans as well as the negative effects of the white storytellers. It just takes time, study, and an attitude correction in order to understand what you are about to read.

Prelude

For centuries, we searched for our ancient past in dead inanimate fossils, graves, and ruins. We have since learned that we each carry a history of our past in our genes, languages, cultures, and religions. They are like the sands of time; that is, they tell how long we've been here and, in worldwide studies, tell approximately where we came from.

Scientists have been analyzing genes from people around the world. (ref. 1, 2, and 3) Their evidence about our origins here on earth should have a huge impact on attitudes about race, religion, and society in general. Why? Their evidence is that:

1. <u>We did not evolve from apes or Neanderthals</u>;
2. <u>We appeared "suddenly" in East Africa about 200,000 years ago</u>; and,
3. <u>Migrations from Africa and continuous genetic mutations have resulted in the diversity of races, languages, and cultures around the world</u>.

The scientists are the first to say that genes and fossils alone cannot tell where, when, or how we got here. However, as their story of our genetic history unfolds, the reader cannot escape the striking similarity with the Hebrew Book of Genesis and the stories of Adam and Eve and Noah's Ark.

These ideas are relatively new to European and Asian cultures; however they have long been part of African oral history and belief. In America, we see everywhere in our black communities slogans such as "Before History, there was Black History" and "African Woman, Mother of Mankind."

Let's take a journey back in time. The clues are faint, but the evidence from genetics, linguistics, archaeology, Genesis, and pre-Egyptian history lead us back to ancient Africa. There, a priestly caste of astronomers, agriculturalists, and stone architects came down the Nile, their legacy perhaps more salient in our languages and religions than in the inscrutable stone monuments they left behind. Their thoughts and values flow repeatedly through our lives like the air we breathe and the water we drink.

They are still among us.

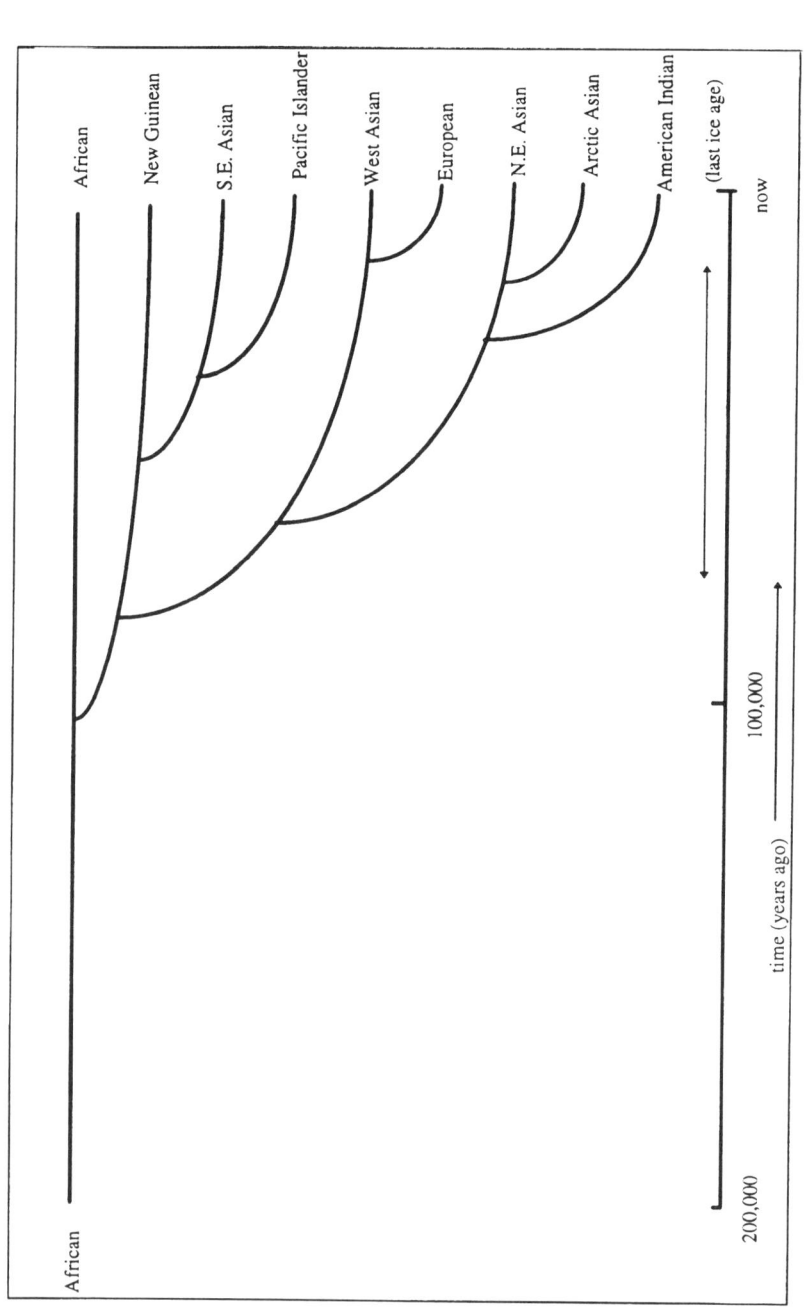

Figure 1. **Human Genetic Family Tree**
(adapted from ref. 2)

Our Common African Genesis

1. Genes

The genetic research starts with the fact that no one is an identical twin of their parents; stated more technically, every new child has small, random, mostly inconsequential, genetic mutations from the parents. These mutations occur more in genes that don't control major bodily functions and physique. There are two sets of genes studied, one that mutates "fast," from generation to generation, and another set that remains almost unchanged within the same families and ethnic groups, but shows differences between groups that have separated for some time. The first set of fast mutating genes are like "genetic clocks" that tell how long we've been here. The second set is used to find the "branches" on the human family tree. <u>Geneticists conclude that we didn't evolve from apes or Neanderthals because our "genetic clocks" simply don't go back that far in time. They conclude that we came from African parents because</u> (1) <u>African have the most mutations and</u> (2) <u>the "branches" on the human family tree don't join to a common "trunk" until we get to the older African genes</u> (ref. 1, 2, & 3).

The genetic human family tree, shown in Figure 1 (adapted from ref. 2), has four major branches, from top to bottom—African, Austronesian, Caucasian, and northeast Asian including descendants of the latter who migrated on to North America and became what we now know as American Indians.

Geneticists draw the family tree differently, giving equal length to all pairs of branches. They do so because we really don't know which races branched off first, second, and so on, and they don't want to artificially bias the appearance of the tree. The author is not so parsimonious; there's no harm in showing the African branch of the tree as the trunk, if that's the case, and showing the first longest branch off that, the next longest off that, and so on. There will always be uncertainty about the early and intermediate branches, who those people were, where they branched off from each other, and which of the present races they looked like, <u>if</u> any. Just don't ignore the trunk.

Note that the European branch is the shortest, most recent branch on the family tree, meaning that these Caucasians are the latest race on earth, the latest mutation of the Mother Race.

However, this is <u>not</u> a tree of evolution; this simply shows <u>when</u> different mutations took place <u>within</u> our species, which corresponds to when different groups separated from each other. We all come from the same gene pool; otherwise we couldn't interbreed. The next chapter deals with this tree as a rough roadmap of the earliest migrations.

The theory of evolution, such as it is, is discussed in chapter 4.

2. Migrations

The family tree takes on more meaning when it's superimposed onto a map of the world as in Figure 2 (adapted from ref. 2). Then, it's easier to imagine how the tree branched knowing where we started and where we ended up. We can now guess the routes of the first migrations, where people split apart, who stayed put, and who kept going. The branches of the human family tree, superimposed on the worldmap, look like the first migrations of modern humans.

The actual routes are, of course, purely speculative, educated guesses at best. Geography and climate should give us some clues; for instance, these initial migrations took place during the last ice age, when much of northern Europe, Asia, and North America were under glaciers, very cold and the oceans were three hundred feet shallower. [It's hard to imagine anyone moving north except to escape nasty southern in-laws.] However, as the glaciers melted and receded north, regions like the Sahara became hotter and drier and northern regions became warmer and wetter. (See chapter 7 for more info on the ice age.) There have been many more migrations since, over the same paths but not in the same directions, so we should not be confused by all present-day populations and languages.

It appears that the first eastern migrations out of Africa could well have been along the coast of the Indian Ocean, often by water, populated the southern coasts of Arabia, Persia, India, Indochina, pushed on to the Philippines, Indonesia, New Guinea, Australia, and the Pacific Islands. [I say "often by water" because by the time these first migrants reached the Pacific Ocean about 40,000 years ago (ref. 2), they must have become accomplished mariners in order to take off across the ocean beyond the horizon. How did they know there was land out there?] We know that the aboriginal people of these areas were a dark-skinned race often called "Negritos" in European references or "small Negros." The longest (oldest) branch of the human family tree outside of Africa is that of the New Guinea Negrito and the Australian Aborigine. Where they split away from the rest of the migrants is uncertain, but the split probably became final when they crossed the fifty miles of open ocean between Indonesia and New Guinea (then connected to Australia) (ref. 6, p. 156 & 160). No genetic trace of these people has been detected in between Africa and New

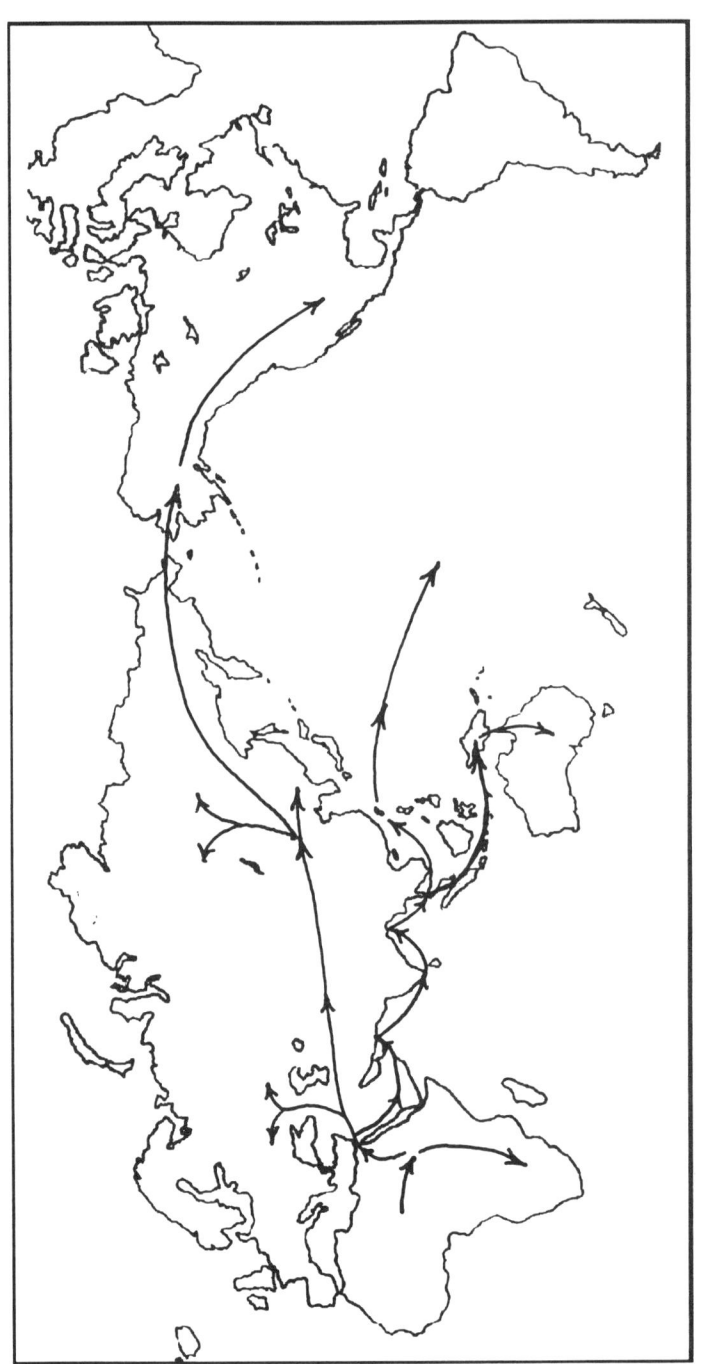

Figure 2. Family Tree of Modern Humans, Superimposed on a Worldmap, with the Branches Pictured as Hypothetical Routes of the Earliest Migrations

Guinea, evidently being "displaced" by later migrations of Caucasians and northern Asians.

It appears that the first northern migrations out of Africa could have gone through the Middle East and into southwestern Asia. The next leg took them east into Asia, where they split into three groups, northeast Asians (Mongolians, Tibetans, Chinese, Koreans, and Japanese), Arctic Asians (Siberians and Eskimos), and the third group that went on across the Bering Strait (then a land connection) into North America (American Indians). The later migrations into Western Europe are believed to have been led by the ancestors of the Lapps of northern Scandinavia and the Basques of northern Spain, because of their genetic age and ancient languages (ref. 4, p. 17 & 63).

Languages, themselves, provide a roadmap into our past, our migrations, conquests, isolation, and absorption. Let's next take a quick look at the world's families of languages and see how they, too, interrelate us.

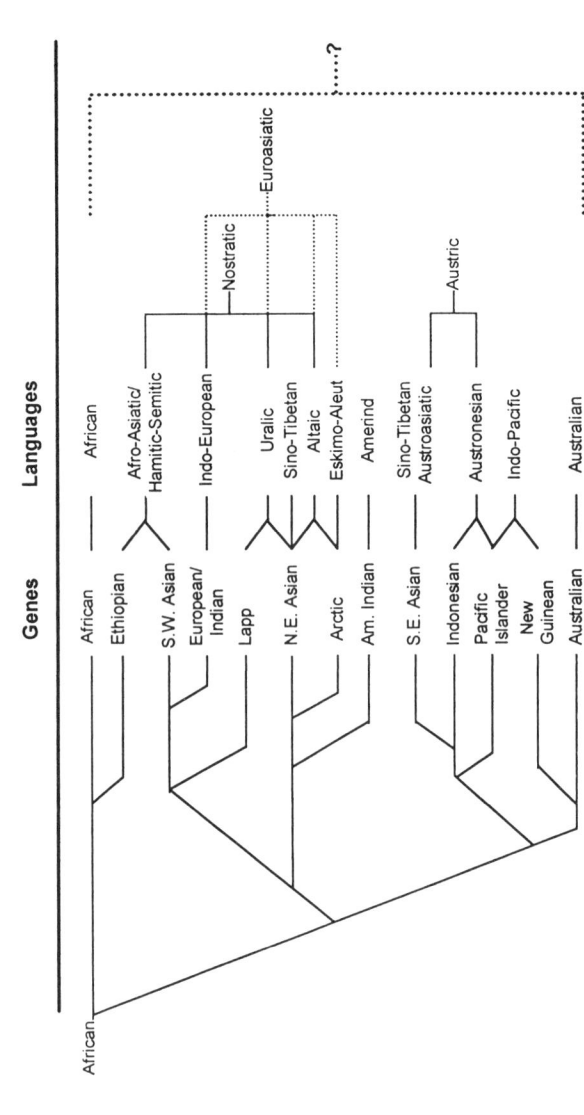

Figure 3. Correlation of Genetic & Language Family Trees (adapted from ref. 2)

3. Languages

Linguists have been constructing a family tree of languages from around the world, attempting to discover their interrelationships and to find the "trunk" of the tree or original language. Unfortunately, so far, languages cannot be traced back more than a few thousand years and certainly not back to a mother tongue. Many old languages died out, most had no written form, and those that did are muted in pictographs and hieroglyphics. Several old languages are still in use today like Aramaic (Syria), Hebrew (Israel), Coptic (Egypt), Geez (Ethiopia), Hausa (Nigeria), and Sanskrit (India), but there are still loose ends. Several other "independent languages" are suspected to have great antiquity because they have no known relationship with any other language, such as Basque (Northern Spain), Mbuti and Khoisan (Africa), Burushaski (Kashmir), Vietnamese and Muong (Vietnam), Japanese, Korean, Ainu (Hokkaido, Japan), Australian Aborigine, and American Indian (ref. 4, p. 2–9). Note that, except for Basque, these are all along the rim of the Indian and Pacific Ocean, the likely route of the earliest migrations out of Africa.

However, the best that linguists can do, so far, is classify about twenty major language families and about fifty lesser ones out of the thousands of spoken tongues around the world (ref. 4, p. x). The existence of one language over a large geographical area is thought to be the result of migrations, either through the spread of farming or military conquest, over the last 7,000 years (ref. 5). The small pockets of isolated independent languages, such as those, above, are regarded as remnants of earlier migrations, over 7,000 years ago (ref. 5). These pockets became isolated or surrounded by the larger language families in the later migrations. Perhaps the genetic family tree is the roadmap that linguists need to trace back further into language origins.

When a language family tree is laid next to the genetic family tree, as shown in Figure 3, they do compare very well between subgroups (ref. 2). That is not surprising to everyone; for example, Charles Darwin, himself, in his treatise on the origin of species (1895) (ref. 38), speculated that as mankind evolved, so would languages. Another famous scholar H. G. Wells, in his *Outline of History* (1920) (ref. 26), speculated that Aryan (European) and Semitic (Hebrew and Arabic) languages were probably proto-Hamitic (African). Not being a linguist, he probably was more influenced by the ancient civilizations along the Nile than by the languages. He also was influenced by some popular theories of his day that spoke of an ancient Neolithic culture that spread from Africa around the world and built the many large stone edifices like the Egyptian pyramids, Stonehenge, and others.

None of this should be too surprising because, for years, languages were named after the perceived race of the speakers or the nation or geographical areas where they were spoken. The correlation is, therefore, built into the names we give people and their languages, not necessarily between genes and languages.

The real surprise comes in chapter 11, where the old Hebrew names are given their original Egyptian translation and meanings; to wit, the names are synonymous with the Hebrew stories in the Books of Moses (ref. 16, p. 143–170). <u>The origins of this language family known as Hamitic-Semitic or Afro-Asiatic then becomes key to our understanding of the African roots of our Western culture and religions.</u>

Next, let's see what the Darwinists have to say about our sojourn on earth.

4. Evolution

What about the Theory of Evolution? Where does it stand in this story about human genes? We didn't evolve from apes or Neanderthals because our "genetic clocks" only go back 200,000 years, not millions of years. We have mutated into different looking people with different skin color, facial features, and types of hair, but these are not changes of species. Our racial differences are shallow, not deep, much to the consternation of some proponents of evolution and to people with a racist agenda. Neanderthals and other two-legged, tool-making creatures preceded us here on earth, as though steps in an evolutionary process, but we still have none of their older genes. We coexisted with Neanderthals for as long as 40,000 years (ref. 3, p. 73), but evidently didn't interbreed with them, if that was even possible.

Geneticists, themselves, are extremely ambiguous about evolution. On one hand they say that "Archaic females do not seem to have contributed . . . genes to the modern people . . . " (ref. 3, p. 73) and, yet, on the other hand, say that " . . . the human DNA clock has ticked steadily for millions of years . . . " (ref. 3, p. 72) simply because they compared it to chimpanzee DNA!? They seem to recognize their contradiction and go to say, "How one human population might have replaced archaic humans without any detectable mixing is still a mystery." (Ref. 3, p. 73). What they are really saying is that they have absolutely no genetic evidence for the evolution of modern humans from any archaic beings, but they still can't give up the concept. Such is the problem with old logical sounding theories; they become part of our language and beliefs and never go away. They become dogma.

The word "evolution" has subsequently grown in use to mean any change with the passage of time, even where there is no evolution of species. The geneticists sink to their lowest when they draw up an imaginary tree of primates, including humans, show Africans evolving from monkeys, and then, crown that tree with white folks, further up the tree (ref. 30)!? Thank you, Mr. Darwin and loyal followers, but no thanks. We cannot discuss evolution in studies of genes that are by definition "neutral" or "inconsequential"; to do so is oxymoronic or simply a contradiction of terms. Please?

Let's see what the diggers of graves and fossils have to say. They've

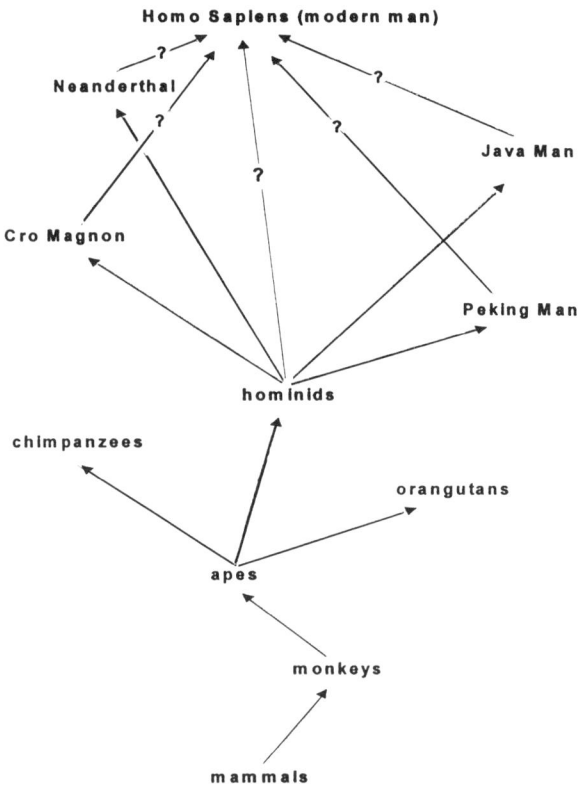

Figure 4. Primate Evolutionary Family Tree

carried the burden of proof for evolution for decades. Geneticists are relatively new to evolution, strange as it may seem, except for their inept writing about it. The old-timers are the anthropologists and archaeologists who examine the ancient remains. They've had several competing theories of evolution, one of which they call the "Out of Africa, Noah's Ark Theory" (ref. 6, p. 128). Surprise? Surprise! Indeed, studies of fossils also supply evidence that modern man originated in Africa, migrated around the world, and displaced earlier populations of Neanderthals.

The perceived evolution of humans and other life forms appears to have happened in steps, during major climatic changes brought on by the several ice ages that have occurred over the past millions of years. This fuels the argument that evolution is an adaptive process, particularly to changes in climate. During these climatic extremes, many species of plants and animals became extinct, while others appeared as if by magic (ref. 6, p. 88–89).

Such transformations of species would not necessarily be considered to be evolutionary were it not for the existence of man and our two-legged "predecessors," the hominids. The evolutionary path to modern man has allegedly taken us from monkeys to apes, to bipedalism (walking on two legs), to increased brain size, to tool-making, etc. (ref. 6, p. 88; ref. 23, p. 460). Therefore, unlike the rest of the plant and animal kingdom, with man there has been more than a simple adaptation to environment. There has been an increasing ability to manipulate or control our habitat, although manifested almost entirely in our own species.

There appear to have been many branches on the primate/hominid/human "family tree," judging from the fossil evidence; that is, over the past millions of years, there were many more varieties of monkeys, apes, and prehuman hominids than there are now. Most of these species died out. Figure 4 is the author's illustration of such a "tree" based upon the concepts of evolution written by proponents of such (derived from ref. 23, p. 448). Which branch of this so-called "family tree," we allegedly came from is not known. There are efforts to recover or synthesize the genetic data from fossils that will better define this "family tree" (ref. 44, p. 225; ref. 45, p. 5).

All of our hominid predecessors between ourselves and chimpanzees died out. That leaves the chimps as our closest living evolutionary relatives . . . with only a 1 percent difference in genetic makeup!! (ref. 6, p. 48). [I have some cousins I didn't think were that close.] Geneticists estimate that we hominids split from the apes around 5 million years ago and progressed to our present form since then.

In spite of all of the fossil evidence and the fanciful explanations about it, there is no genetic model of evolution. There is just the very strong belief and hope that there is one, yet to be discovered. Genetically, we modern humans go back to a common mother in Africa about 200,000 years ago. Period.

We shouldn't leave evolution with a totally negative attitude. There are scientists with more carefully stated views of the history of the earth and its fossil contents; for instance:

W. Lee Stokes: " . . . if rock layers [in the ground] are progressively older downward, so are their fossil contents—the older the rocks, the older the fossils. . . . Once the fossils are arranged in their . . . natural order [in time], . . . it is obvious that they show an orderly progression from simple to complex, an order that is best explained by organic evolution" (Stokes, W. Lee, *Essentials of Earth History,* fourth edition, Prentice-Hall, Inc., Englewood Cliffs, N.J. 1982) (ref. 23, p. 110).

Figure 5, from Mr. Stokes's book, illustrates his views (ref. 23, p. 106 & p. 109).

Having gone into evolution for a short explanation and critique, let's next talk about Adam and Eve and old Noah and his ark. Then, let's look at the close of the last ice age and try to imagine what effects the release of water from the melting ice were on the landscape and on the people who survived the ensuing floods. Then, we will look into the differences between genetics and the Book of Genesis and try to understand why they differ. Eventually, we will try to see what happened to the Hamitic/African "trunk" of our human family tree and why it was "glossed over" in the Judeo-Christian version of our ancestors.

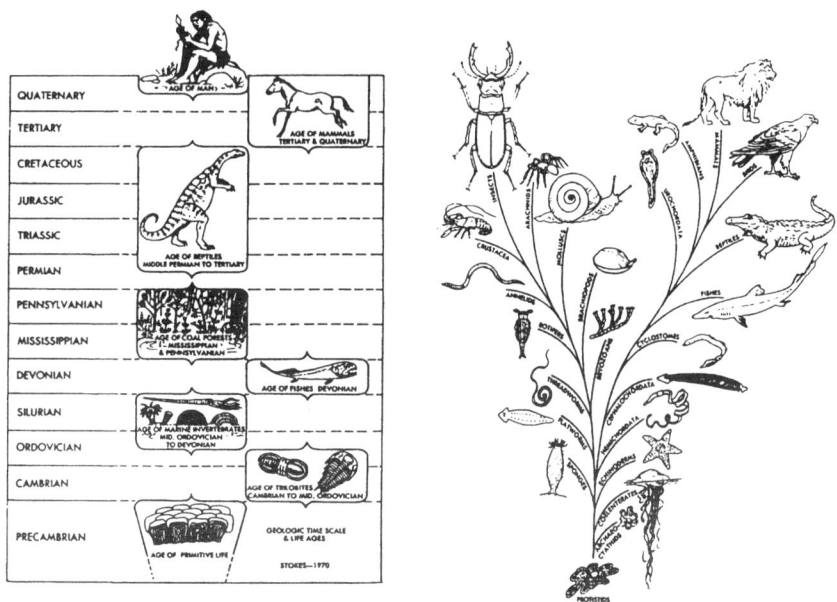

Figure 5. **Origins of Organic Evolution**

a. Simple-to-Complex
(layers into the ground)

b. Resultant Family Tree of Animals
(family tree of animals, simple-to-complex)

5. Adam and Eve

What about Adam and Eve in this genetic story? <u>Geneticists trace our ancestry back to a common mother</u> (Eve?) <u>in Africa,</u> because one set of genes (mitochondrial) that they study is passed from mothers to their children with no mixing with those of the fathers (ref. 1, 2, & 3). [The male mitochondrial genes are in the tail of the sperm and are subsequently lost at conception.] The trace back to one common mother is just a statistical result, because other mothers of that time and until ours had lines of descendants that were missing female offspring, not an uncommon occurrence. Population geneticists estimate that as may as 10,000 people lived at the time of our African Eve (ref. 3, p. 70). After all where did the sons of Adam and Eve find their wives? And, who were the "sons of God" and the "daughters of earth" whose fraternization displeased God? Let's look further.

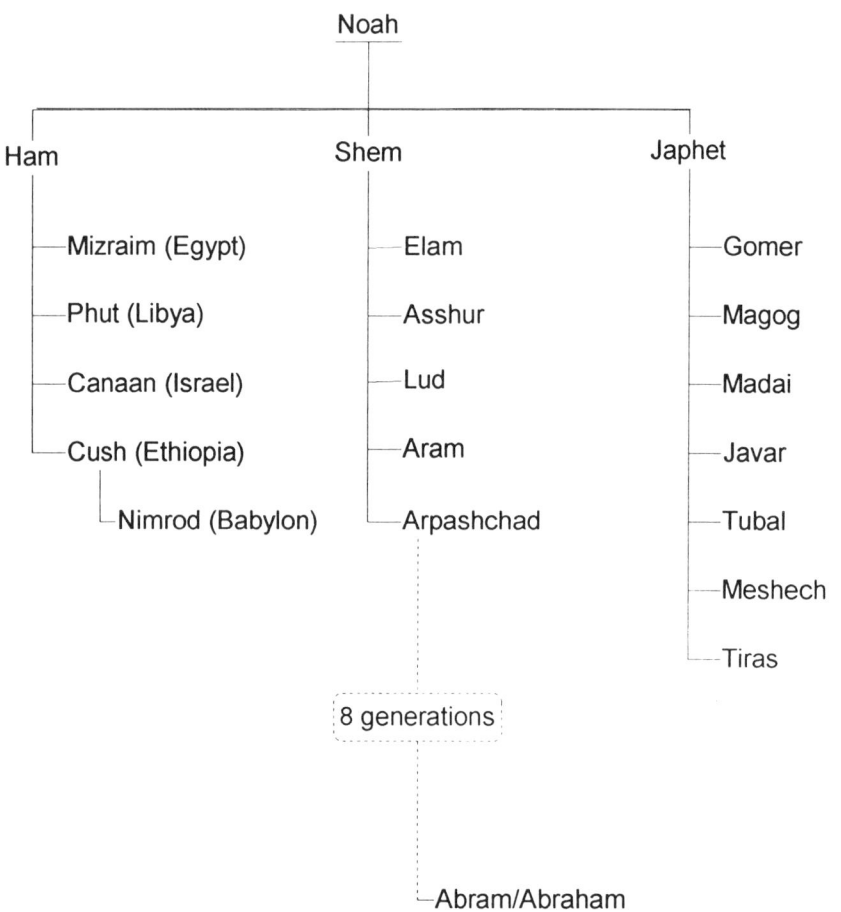

Figure 6.
The Table of Nations,
The Family of Noah
(Genesis 10)

6. Noah's Ark

The most exciting analogy with the genetic story is that of Noah and his sons Ham, Japhet, and Shem and their migrations after the flood (Gen. 10). The sons of Ham are all identified with ancient biblical kingdoms that are recognizable today—Cush, now known as Ethiopia and Sudan, Mizraim, now known as Egypt, Phut, now known as Libya, and Canaan, now known as Israel. Cush's son, Nimrod, was the biblical founder of Babylon, now Iraq. The descendants of Noah, from Genesis 10, are shown in Figure 6. The migrations out of Africa, north by Canaan and east by Nimrod, are very much like the earliest migrations in the genetic story. Archaeology has always placed the first civilizations in the area of Egypt, Canaan, Syria, and Iraq. The genetic and biblical evidence swings this around the crescent towards Ethiopia, the ancient home of the black Hamites and Cushites.

<u>Without evidence to the contrary, the sons of Ham appear to be the first builders of cities, industry, and trade.</u> Such technology is not the product of nomadic shepherds. When Abraham came out of Chaldea, Canaan was long established. Likewise, when Moses led his flock out of Egypt, Jerusalem, Jericho, and other cities had already been built by the Hamitic descendants of Canaan.

Not to slight the other sons of Noah, but Genesis provides little information. Abraham, Moses, and Mary came from the line of Shem and that's all we are told. This essay eventually gets to the puzzle of when, where, and why the "line of Shem" appropriated the Hamitic cultures and made history and religion look like it started with the Semites and no one else.

Next, let's look at the earth's climate during and at the end of our earliest migrations and preceding the Book of Genesis. In particular, let's examine the end of the last ice age and think about the rains and floods that must have occurred as the ice melted, rains increased, and the oceans rose. This just preceded the historical age and left its mark on our mythologies of ancient times.

7. The Ice Age and Early Man

Modern man is a survivor of the last ice age. As the earth cooled and the glaciers spread southward, we, modern men and women, migrated from Africa and spread around the world (ref. 6, p. 156). Later, as the earth warmed and the glaciers receded northward, many plants and animals died out, particularly the large mammals, while we flourished everywhere on earth. The so-called stone age of mankind ended about the same time as the ice age (ref. 6, p. 159).

The last ice age began about 75,000 years ago with a drop or a steady decline of the earth's temperature (ref. 6, p. 160). Glaciers grew out of the northern areas and spread over large parts of North America and Europe and a smaller part of Asia. Isolated glaciers also grew at high elevations everywhere including New Guinea and Africa. The cooling of the earth and the size of the glaciers reached a maximum about 20,000 years ago (ref. 6, p. 160). Figure 7 is a world map and overlay that illustrates the maximum extent of the glaciation (derived from ref. 23, p. 404, ref. 28, p. H2, & ref. 6, p. 158 & 160, ref. 46, p. 40, ref. 43, Ice Ages).

Glaciers then covered almost one-third of the earth's surface, mostly in the Northern Hemisphere, reached thicknesses of several thousand feet, and the oceans dropped as much as 300 feet (ref. 23, p. 419). The climatic zones between the equator and the Arctic didn't disappear; they were compressed southward into narrower belts around the world (ref. 23, p. 415). Regions like the Sahara experienced rain because the rainbelts shifted south from Europe. Otherwise, rain and snowfall declined worldwide because 1) more water was locked up in the ice and 2) lower temperatures resulted in less evaporation from the oceans. Lakes and rivers grew in now dry areas. For example, the Dead Sea in Israel, now 1,100 feet below sealevel, was at one time 1,200 feet higher (ref. 23, p. 415). The Caspian and Black Seas in southeastern Europe formed one giant lake just to the north of the Middle East (ref. 28, p. H2).

From 15,000 years ago on, the earth warmed up, the glaciers receded northward as they melted, and the oceans rose in height. Big lakes were formed by the water from the melting glaciers, locked in their natural rock dams of earth and rock, that had been pushed ahead of the ice (ref. 23, p. 415). About 13,000 years ago, the warming trend increased dramatically. This resulted in faster melting of the ice, more rain, faster rise of ocean lev-

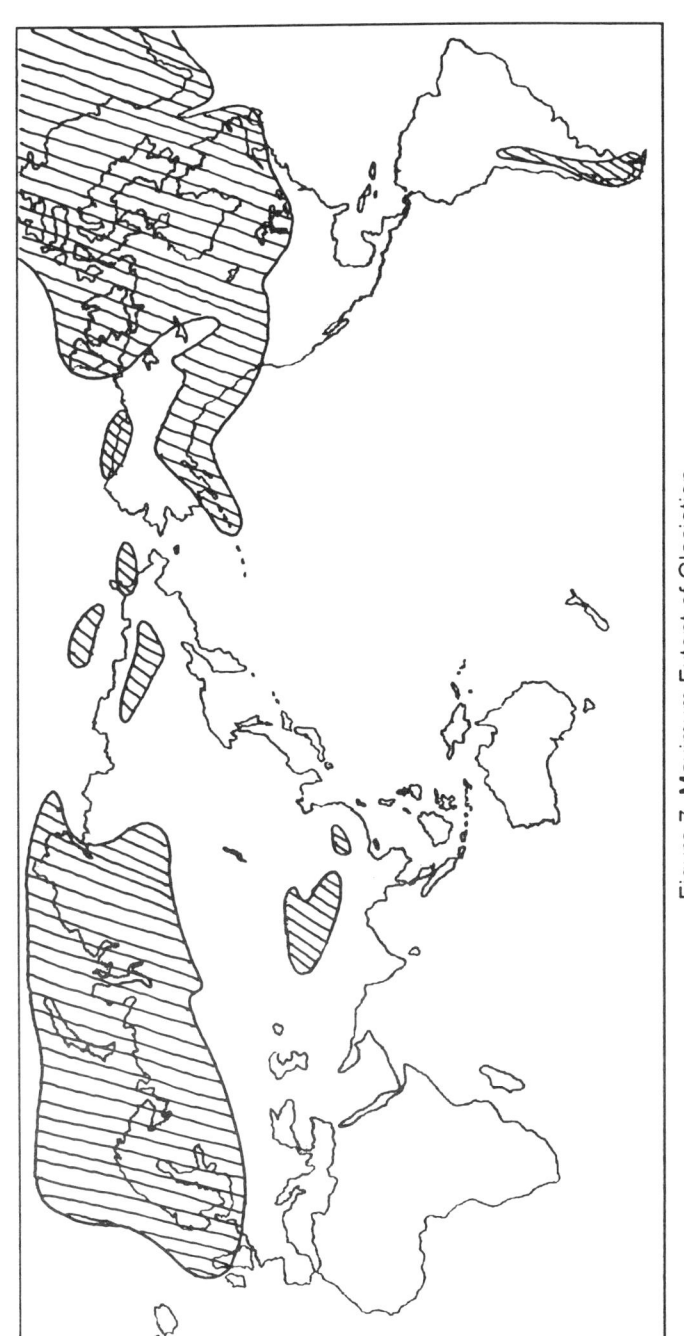

Figure 7. Maximum Extent of Glaciation, Approximately 20,000 Years Ago (adapted from ref. 6, 23, & 28)

els and, undoubtedly, extensive flooding (ref. 23, p. 413, ref. 6, p. 167). The warming trend peaked around 11,500 years ago (ref. 23, p. 419), the last convulsive gasp of the ice age.

From that time on, for about 4,000 to 5,000 years, many species of animals died out everywhere, particularly the large mammals (ref. 23, p. 427). Whatever the reason for these mass extinctions, men started farming and domesticated some of the remaining smaller animals. Then, our populations exploded and we entered the "age of mankind" (ref. 6, p. 169). These events were experienced by modern man after we had migrated just about everywhere on earth. The rains and floods must have left an indelible mark on man's memory, filling the conversations with the power of nature and the resultant tragedies for mankind. [Imagine, if you will, the melting of the ice in Antarctica in our time. The oceans would rise to the chin of the Statue of Liberty! (derived from ref. 23, p. 419).]

All of this time, from 70,000 years go to historical times, man was migrating like the rest of the flora and fauna, staying on the move, trying to stay alive. Never was the whole world under water, but there was catastrophic flooding at different times at different places due to the melting ice and increased rain. Huge areas of the seashore were inundated as the oceans rose three hundred feet. Stories of such floods abound in our legends and religious mythologies. The men who survived were obviously blessed and the men who perished were obviously cursed.

The Neanderthals lived close to the glaciers during the ice age and perished for unknown reasons. It is easy to imagine that these strange, near-human, beings might have been those mentioned in Genesis that God "needed" to eradicate from the earth and perhaps many died in the floods. It's strange to imagine that these near-human creatures may have been mentioned in the Bible.

Next, let's compare Genesis to genetics and look at some simple but major differences. Such differences may be due to who wrote the Books of Moses and what were their intentions.

8. Genesis versus Genetics

Genesis and genetics don't agree completely. There is a huge difference in their calendars. The Hebrew calendar goes back only 6,000 years, while the genetic calendar goes back 160,000 years or more. Both are well founded within their own references. So, why this difference and is it important?

Perhaps Adam and Eve were not the first, but merely the first with a family tree that was preserved through the ages. There are three other groups of people that are mentioned in Genesis that are not explained; these are 1) Adam and Eve's in-laws, 2) the "giants of the earth" who existed before and after the flood, and 3) the "sons of God" and the "daughters of earth" whose relationships created more people. The "giants of the earth" are mentioned in Genesis 6.4, before the flood and in Numbers 13.32 and 13.33, long after the flood, when Moses' spies went into Canaan. However faint the clues, Genesis does mention these ancient people, not related to Adam and Eve, some of whom survived the flood.

It should not be surprising that there were those whose history did not make it into the Book of Genesis. Genesis was told by Hebrews, about Hebrews, for Hebrews, and not written down until 1,000 B.C. under King Saul's reign, to 600 B.C. during the Babylonian exile (ref. 7, p. 271). After the Tower of Babel, that history was limited to the descendants of Abraham, the patriarch of the modern Jewish people (ref. 7, p. 35). Stories of Creation abound in all cultures and stories of a great flood were told around the Middle East as early as 4,000 B.C. by people outside the deluge (ref. 7, p. 31). It then seems that the Hebrews didn't own the copyright to either story.

We can look for differences between Genesis and genetics in the following:

1. the "genetic clock" or the Hebrew calendar or both are wrong, or
2. if the two time records are both correct, then **the Hebrews were not the first and Genesis is an incomplete history/mythology of early mankind.**

The similarity between the two is so good that the stories of Creation and Noah's Ark should not be dismissed simply as allegories or myth. This will be discussed again after we delve into a few more topics.

9. Pre-Egyptian History

Many students of pre-Egyptian history place the earliest civilization in the land of Cush, the same area that geneticists say is our original homeland. The Cushite language has been found in ruins from Africa to Babylon to India. The first phonetic alphabet is credited to the Phoenicians, originally Canaanites. Stephanus of Byzantium wrote, "Ethiopia was the first established country on earth and the Ethiopians were the first to set up the worship of the gods and to establish laws" (ref. unknown, but quoted in ref. 15, p. 8–9). The Greek philosopher Homer said that the Cushites were the most just of all men—the favorites of the gods (ref. 12, p. 221). The Greek chronicler, Herodotus, describes them as the tallest, the most beautiful, and long-lived of the human race (ref. 9, p. 161). A timeless tribute to the Cushites is the names of the star constellations after King Cepheus, his wife, Cassiopeia, and their daughter, Andromeda, an ancient royal Cush family (ref. 15, p. 12). Herodotus also wrote that (1) the Greeks got their gods' names from the Egyptians (ref. 9, p. 87). Alexander the Great made a dangerous trek to Amman, the home of the Egyptian gods, in order to verify his bloodline with them (ref. 17, p. 108; ref. 46, p. 151).

Egypt was never young. The oldest pyramids contain technology from an older advanced culture. Egyptian civilization most likely came down the Nile from Ethiopia. As stated in the preface, recent archaeological investigations have uncovered sophisticated tools deep within Africa that are about 90,000 years old (ref. 24, p. 378).

Egypt established colonies or outposts in Canaan (ref. 39, p. 215), Greece (ref. 17, p. 59), and Colchis (ref. 9, ref. 17, p. 69) (southwestern Russia) early in her history.

After Alexander the Great conquered (actually, liberated) Egypt, his successor, Ptolemy, founded a new city, Alexandria. He founded a major university with research facilities and a monumental library. Written works of the ancient world were collected and copied for study (ref. 46, p. 153). The soon-to-become great thinkers of Greece went to this school and taught at this institution. We have been led to believe that the Greeks were "Hellenizing" Egypt; the opposite may be the truth.

Several turn-of-the-century European scholars found evidence of Cushitic predominance at the dawn of history and left us some colorful quotations that summarize their findings:

Sir Henry Rawlinson: "A laborious study of the primitive language of Chaldea led him [his brother, Henry] to the conviction that the dominant race in Babylonia at the earliest time to which the monuments reach back was Kushitic" (George Rawlinson, *The Origins of Nations,* pp. 212–214, New York, Scribner, Welford & Armstrong, 1878) (ref. 18).

J. G. R. Forlong: "it was undoubtedly Kushites who rendered possible the Aryan advance, and who played the part of civilizing Rome, thousands of years before Roma's birth. It was their vast mythologies and strange legends that passed, as Lord Bacon wrote, 'like light air into the flutes of Grecians, there to be modulated as best suited Grecian fancies.' Indeed, it is manifest from many old writings, that it was their tales, myths, traditions, and histories that lay at the base of the Western World's thought and legendary lore. These so impressed all subsequent races, and entered so deeply and minutely into all Aryan mythologies, that many writers now think Aryans can only claim to have added to the superstructure and complexion of Ethiopian myths and mythical history, . . . and let us remember that active Aryan life and mythologies began at least 3000 B.C. when high Asia . . . becoming too cramped for this race . . . was pressing southward to India and Ariana and to the west generally. Then and there must Aryans have met with Ethiopian civilization, as did Semites, when these began to group themselves into nations about a thousand years later, or say, 2000 B.C. They were all builders on old Kushite Foundations" (J. G. R. Forlong, *Rivers of Life,* vol. 2, pp. 403–404, London, Bernard Quaritch, 1883) (ref. 19).

Charles Seignobos: "It is within the limits of Asia and Africa that the first civilized people had their development—the Egyptians in the Nile Valley, and the Chaldeans in the plains of the Euphrates. They were people of sedentary and peaceful pursuits. Their skin was dark, the hair short and thick, the lips strong. Nobody knows their origins with exactness, and are not agreed on the name to call them (some terming them Kushites, others Hamites). Later, between the twentieth and twenty-fifth centuries B.C. came bands of martial shepherds who had spread all over Europe and the west of Asia—the Aryans and Semites" (C. Seignobos, *History of Ancient Civilization,* p. 17, London, T. Fisher Unwin, 1907) (ref. 20).

A most remarkable book was published in the 1920s by a most remarkable lady, Drusilla Dunjee Houston, a black American, in the middle of Oklahoma, USA. She tracked the history of the human race that parallels the genetic story. From the *Wonderful Ethiopians of the Ancient Cushite Empire,* we offer some wonderful quotations:

"The Cushite race, its institutions, customs, laws and ideals were the foundation upon which our modern culture was laid" (p. 11). . . .

"General history informs us that when the curtain of history was lifted, the civilization of Egypt was hoary with age" (p. 16). . . .

"Earnest and conscientious students, seeking the facts about ancient Ethiopia, find but scanty and unsatisfactory references in modern books. Going back to ancient records we find voluminous testimony" (p. 49). . . .

"Renan asserts that Egypt had no infancy, no archaic period, because her first colonists were civilized in Ethiopia" (p. 93). . . .

"Bunsen believed that the time preceding Menes [~4,000 B.C.] was greater than since. Lepius says, under the Fourth Dynasty, six thousand years ago, the nation had approached the highest development at which we find her, of which the ruins still bear witness. The admirable system of monumental writings showed its highest perfection in the oldest ruins. This certainly indicated a long period of development" (p. 69). . . .

"Egyptian civilization was highest at its first appearance showing that they drew from a fountain higher than themselves" (p. 80). . . .

"1700 B.C. finds Egypt invaded and conquered. Egypt had broken into two really separate kingdoms. This enfeebled the country for the conquest of the Hyksos. During their stay, the native princes at the south maintained themselves. 2080—1525 B.C. these Shepherd Kings ruled over Egypt. They were a barbaric and nomadic race from Asia which destroyed the temples and left no monuments standing in Egypt. . . . Egypt entered into the darkest period of her history. . . . The Shepherds were expelled from Egypt by Aahmes [a prince from the south], . . . Under Aahmes Egypt again became supreme. The decayed and ruined temples were restored to their ancient richness and splendor. In a few years she had regained what had been lost in the five centuries of rule under the Hyksos. . . . Aahmes founded an empire that lasted 1,500 years, a period rich in its records of history and growth for Egypt" (p. 98–99). . . .

"Tut-ankh-amen [King Tut] was born 1350 B.C., long before the days of Athens and Rome (p. 103). . . . His name ended in Amen, the black god of the Sudan and Egypt. With his name the Egyptians began and ended their prayers. We of the Christian world, through Hebrews, have appropriated it and use the title of the great Amen at the close of our petitions" (p. 104).

"Semites made no showings of culture until the rise of the half barbarous Assyria, which copied its arts and sciences from Cushite Chaldea.

The Hebrews learned agriculture and building from the Hamitic race of Canaan" (p. 54). . . .

"The Phoenicians in the time of Christ called themselves Ethiopians. The Scriptures and ancient records called the Samaritans Cushites" (p. 22) (Drusilla Dunjee Houston, *Wonderful Ethiopians of the Ancient Cushite Empire,* Black Classic Press, Baltimore, 1985) (ref. 17).

A more recent scholar, a linguist, offers the following views about the origins and migrations of the Indo-Europeans [a.k.a.: Caucasians, Aryans, or whites], based upon linguistics and history, which is relevant to who was where first and so on:

Kenneth Katzner: "It would appear that the Indo-Europeans lived in a cold northern region, that it was not near water, but among forests, . . . and that among metals, they knew only copper."

"The general consensus is that the original Indo-European civilization developed somewhere in Eastern Europe about 3000 B.C. About 2,500 B.C., it broke up; the people left their homeland and migrated in many directions. Some moved to Greece, others made their way into Italy . . . while still another branch crossed Iran and Afghanistan and eventually reached India. Wherever they settled, the Indo-Europeans appear to have overcome the existing populations and imposed their language upon them" (Kenneth Katzner, *The Languages of the World,* London & New York, Routledge, 1977, rev. 1986) (ref. 4, p. 10).

For a succinct historical footnote, let's hear how Herodotus described the Egyptians he visited around 450 B.C. Many scholars and would-be-scholars have claimed that the Egyptians were Caucasian. Herodotus' simple observations have not always been well received by scholars because he lacked the 100 percent hindsight that a couple of thousand years gives scholarship; you see, Herodotus had a unique problem—he was there!

Herodotus (on the Egyptians): " . . . they are black-skinned and have woolly hair" (Herodotus, *The Histories,* translated by Aubrey de Selincourt, (1955), revised by John Marincola, Penguin Books USA, New York (1996) (ref. 9, p. 121).

A modern author, Dr. Donald B. Redford, an Egyptologist from the University of Toronto, has written a comprehensive book, "Egypt, Canaan, and Israel in Ancient Times" (ref. 37) that uses the history and archaeology of that region to address some of the same issues as this essay. We are going to provide some of his words, here, in order to shed some light on the issues, hopefully without taking too much out of context and not distorting the themes and opinions that he offers.

"... the general demographic flow that seems to be charted in the distribution and sequence over time of Neolithic sites in the Nile Valley would seem to be from south to north. Consonant with this pattern is the apparent African connection of so many of the early traits of Neolithic culture in Egypt . . . " (p. 7).

However, the general trend of Dr. Redford's thesis is the influence of Asia on Egypt, partly because of a 3,000-year gap in Egyptian historical and archaeological evidence, between 9,000 B.C. and 6,000 B.C., the period between the older Stone Age and the agricultural and urbanized Neolithic Age. There is more or less continual flow of such evidence from the regions of Mesopotamia and Canaan (p. 5). He summarizes this issue as follows:

". . . there can be no questioning the fact that the Gerzean [an Egyptian period from 3400 B.C. to 3,050 B.C.] displays numerous cultural features that are not the products of autochthonoous [native] development, but which have all the earmarks of having been introduced from the outside suddenly" (p. 17).

". . . in the Tigris-Euphrates Valley and in southwest Iran, the cultural evolution that produced these forms and motifs can be traced back over centuries of indigenous development, whereas in Egypt there are no antecedents. Few would dispute, therefore, the obvious conclusion that we are dealing with the comparatively sudden importation into Egypt of ideas and products native to Mesopotamia" (p. 17–18).

Dr. Redford's conclusions are evidently based upon archaeological data and could change with the next "find."* Whereas the archaeological data is not biased by its original owners, it does represent a kind of "random sampling." To reach cause-and-effect conclusions, "antecedent, therefore because of," seems difficult with this kind of evidence. Now, add a 3,000 year gap in the data?

Herodotus reported Egyptian Pharaohs spanning 340 generations, about 11,340 years (ref. 9, p. 138). How long did it take to learn how to build the pyramid of Zoser in 2,650 B.C. (ref. 45, p. 95)? How long did it take to develop a detailed zodiac, a twelve-month calendar, and astronomical observations unrivaled until the nineteenth century? The average person doesn't know how much trial and error goes into science and

---

*Case in Point: *USA Today,* Dec. 16, 1998, Subject: "Earliest Known Writing Found in Southern Egypt; Estimated Date: 3300–3200 BC."

engineering and how hard it is to transfer in any age, more so in ancient times.

Dr. Redford does go into other forms of evidence, attempts to sift through it all, and make sense of it. He goes into the same time period and people that this author finds interesting and relevant to our times, namely the Egyptians, the Hyksos, the Canaanites, and our spiritual forefathers, the Hebrews.

"Josephus himself, speaking as a Jew, refers to the Hyksos as 'our ancestors,' a curious half-truth which we will analyze later" (p. 99).

"Of great consequence to Egypt was the appearance in the north of two non-Semitic speaking ethnic groups who were to threaten the security and independence of the entire Levant [region around Canaan]. . . . Although their folklore and traditions maintain a disconcerting silence on the subject of their origins, the language of the Hittites, deciphered early in the present century, belongs to the Anatolian [region in Turkey] branch of the great Indo-European family; . . . Presumably in keeping with what is known of the place and origins and movements of other Indo-European speaking peoples, the Hittite ancestors were originally at home in the southern steppes of Asia, and only emerged into the Anatolian [Turkey] plateau in the mid-to-late third millennium [B.C.]. . . . From shadowy beginnings the Hittites emerge into the full light of documented history during the later years of the Hyksos occupation of Egypt" (p. 132).

"We know, . . . that by the end of the fifteenth century [B.C.] a surprisingly large number of the ruling families in the towns of Palestine and central Syria display Indo-European names despite the fact that their bearers spoke 'Canaanite' " (p. 137).

"Nearly every major town in Palestine and southern Syria is found, upon excavation, to have undergone a violent destruction sometime after the close of MBIIC [middle bronze age]—that is, the cultural phase roughly contemporary with the last stage in the Hyksos occupation of Egypt" (p. 138).

The Egyptians of Ahmose's time were notoriously inept when it came to laying siege to, or assaulting, a fortified city" (p. 138–139).

"But even a cursory reading of this account [the Hebrew takeover of Canaan] is bound to excite suspicion. Cities with massive fortifications fall easily to rustic nomads fresh off the desert, a feat Pharaoh's armies had great difficulty in accomplishing" (p. 264).

"A detailed comparison of the Hebrew takeover of Palestine with extra-biblical evidence totally discredits the former. . . . also Egyptian control over Canaan and the very cities Joshua is supposed to have taken scarcely wavered during the Late Bronze Age" (p. 264).

"Egyptian Nilotic society had, since the dawn of time, given practical and moral priority to sedentary life and poured contempt on the uncontrolled movement of people" (p. 271).

With regard to the Table of Nations and the sons of Ham in Genesis 10, Dr. Redford offers this quizzical remark: "Last comes Canaan, a faint reminder of the erstwhile [past] extent of Nilotic influence" (p. 405). [Even a Caucasian author has to acknowledge the extent of early Hamitic or Nilotic influence. But, it stops there. Even though the 600 B.C. Semitic authors of Genesis acknowledge the Hamitic origins of the Canaanites, Dr. Redford does not see fit to follow this clue and find out where the Semitic invaders got their culture or trust their own rendition of this story. Once again, we have history starting with the white race, latecomers, at best.]

"There is only **one** chain of *historical* events that can accommodate this late tradition, [the Hebrew stay in Egypt and the Exodus], and that is the Hyksos descent and occupation of Egypt"* (p. 412).

"In sum, therefore, we may state that the memory of the Hyksos expulsion did indeed live on the folklore of the Canaanite population of the southern Levant [Israel]. The exact details were understandably blurred and subconsciously modified over time, for the purpose of 'face-saving' " (p. 413).

"Along the coast north of modern Haifa, and extending as far as Arvad, did the original Canaanite population maintain itself inviolate in the age-old city-states of Tyre, Sidon, Beirut, Byblos, and Arvad. These coastal Canaanites, now appearing under the rubic 'Phoenician,' were more than ever oriented toward the sea, . . . " (p. 299).

"The reputation of Egypt for 'metaphysical' inquiry into imponderables, **which brought many a Greek of the seventh and sixth centuries to the feet of an Egyptian priest,** vanished in the fifth and fourth, as Greek admiration gave way to contempt.* Similarly, the heights achieved by the Hebrew prophets before the Exile dwarf the attainments in the same spheres of the restrictive, ritual-conscious community of the Second Temple" (p. 470).

" . . . Amun [ancient Cushite god] and Yahweh [the god of Moses] had failed. Both Egypt and Western Asia proved powerless to withstand the

---

*Underlines and bold type are this author's, not Dr. Redford's

onslaught of Greek and Roman arms on the battlefield, and retreated perforce in the realm of ideas as well" (p. 471).

To paraphrase Dr. Redford's next comments, under the threat of foreign powers, the Egyptians' and Hebrews' religions became nationalistic, but as the people watched their nations crumble, they turned more and more to the concept of personal salvation and the worship of Christ-like savior figures. Now follows Dr. Redford's last remarks from his large treatise, a clever way to close his book and tie it ito the present.

"The effects are still with us. The saving grace of an Isis [an Egyptian savior figure], a Christ, or Mithra [a Hittite savior figure] triumphed everywhere in the Mediterranean world of the universal empire of Rome, and in a transmuted state has even descended to the twentieth century" (p. 471) (Donald B. Redford, *Egypt, Canaan, and Israel in Ancient Times,* Princeton University Press, 1992) (ref. 37).

This ends the lengthy quotations of others who have delved into this pre-historical era. It is difficult to cull through all of the diverse opinions about whether civilization started in Caucasia or Africa because of the prevailing prejudices and the inundation of Caucasian-led literature on Semitic predominance in religion and Greek predominance in European history. However, there emerges from these other studies, a general acknowledgment of the early "existence" of the Hamitic, Cushitic, or Nilotic people and civilization in Sumeria, Canaan, and Egypt, even though there is not agreement on their contributions to the later Semitic and Caucasian civilizations. While many question the historical accuracy of the Books of Moses, no one effectively challenges the Table of Nations (Gen.10) and the early placement of the civilizations of the sons of Ham.

Whereas we know nothing about the prior history and culture of the invading sons of Japhet and Shem (Caucasians and Semites) from the north, we do know that our knowledge of their culture begins *only* when they encounter the Hamitic peoples of Sumeria, Canaan, and Egypt. They adopted the language, religions, and culture of the Hamitic peoples whose lands they invaded and their history begins then and there. Period.

For a final tribute to the Hamitic progenitors of civilization, we offer Figure 8, The Ancient Cushite Zodiac (ref. 13, 16, 25), and Figure 14 (p.66), estimated by some researchers to be over 10,000 years old. The Zodiac of Dendura (ref. 11, 25), depicts the annual life on the Nile, in particular the prediction of the annual flooding and the dry season in between.

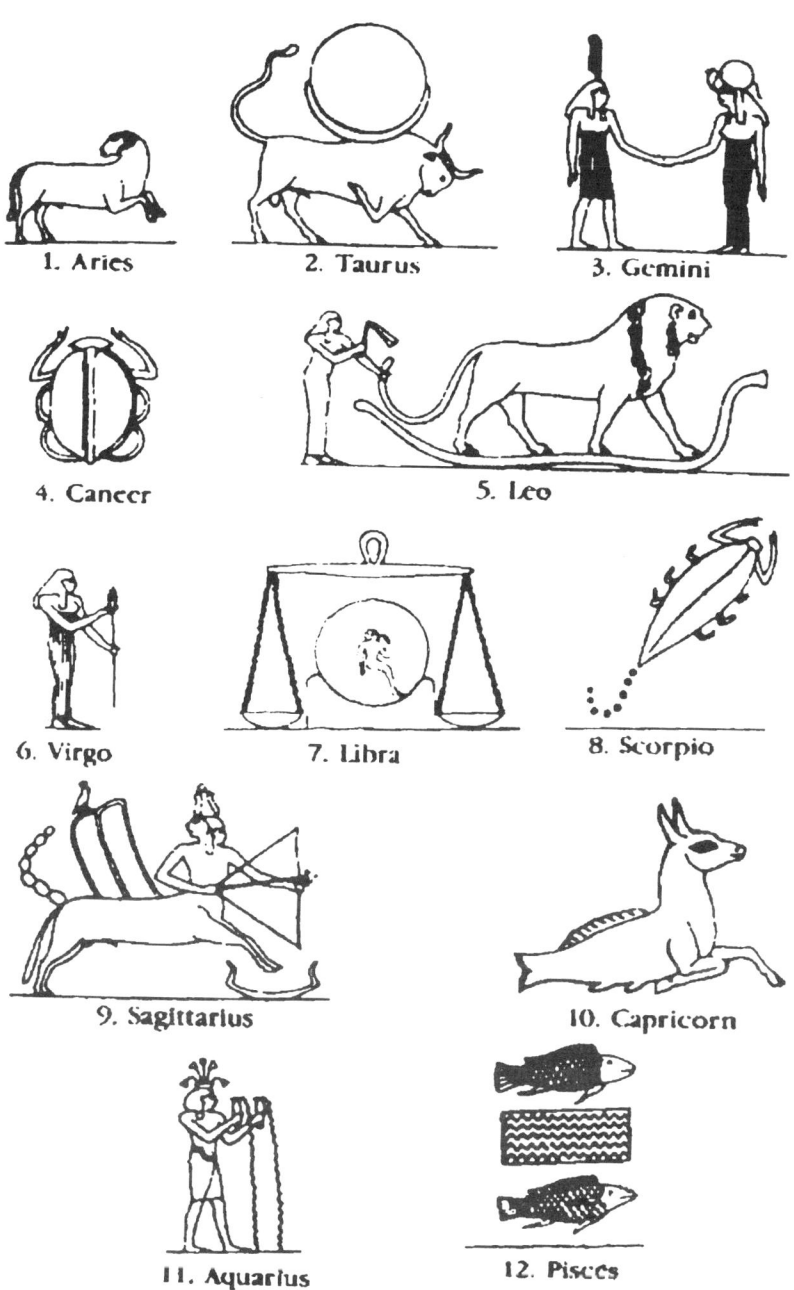

Figure 8. **Ancient Cushite Zodiac**

10. The Early Hebrews and the Cushites*

When Abram/Abraham and the early Hebrews migrated south into Canaan and Egypt, they had few trappings of civilization; they were nomadic shepherds. The Cushite civilizations were already thousands of years old as evidenced in their cities, architecture, agriculture, astronomy, calendar, and, yes, even their religions. The Hebrews were to absorb such things and other cultural amenities from the older race with its "heroes of old, the men of renown" (Gen. 6.4).

The Hebrews have passed on to us the history and religion of their forefathers. However, throughout their stories are the myths, legends, history, and language of those who taught them, the Canaanites, the Egyptians, the Midianites, and the Sumerians. Even with that, the Egyptians despised the Hebrews because they were lowly shepherds (Gen. 46:31). Moses spent years training as an Egyptian priest and/or he spent years with his Midianite (Cushite) wife's priest family before he became the great law-giver of his people (Exod. 2 &3). The Exodus and Conquest by the Hebrews reenact the ancient Song of the Sea from Cushite myths of Creation. The appearance of God in a storm of fire, smoke, and quakes is straight out of the myths and legends of the Cushite god Ba'l, the same as the Greeks' Kronos and the Romans' Saturn, the same as the biblical Nimrod, a son of Cush. The covenant of Moses on Mount Sinai is a series of civil and hygienic codes from the older, more advanced, civilizations of Egypt, Midia, Canaan, and Sumeria.

The Hebrews were "catching up," so to speak. They were, quite naturally, confused and frustrated by the influx of foreign culture, laws, and religion and vacillated and complained. Moses was criticized by his sister for being married to a Midianite (Cushite) woman. Yet, Moses was forced to ask his brother-in-law to guide them through the wilderness. He also asked his father-in-law for advice on managing his large organization, got it, and implemented it—the twelve tribes of Israel with a special financial tribe, the Levites. [It seems that the Israelites were saved by Moses' Midianite relatives, more than from outside help.]

Oddly, two of the three annual festivals mandated to the Hebrews through Moses on Mount Sinai, the Feast of Harvest (Shavout) and the

*Material on Canaan adapted from ref. 29.

Feast of Ingathering (Succoth) (Exodus 23.14), were ancient Cushite farming celebrations. The Hebrews, however, were decades away from becoming farmers. In post-biblical times, Shavout was given another meaning by turning it into a celebration of Moses receiving The Law on Mount Sinai. The Christian holidays of Easter and Pentecost relate back to the Hebrews' Passover and Shavout, but were eventually separated on the calendar in order to establish separate identities. Christmas was moved to December 25 in order to coincide with, replace, or take over the exuberant ancient Cushite celebration of the Winter Solstice, the annual "rebirth of the sun."

Clearly, the Hebrew culture and religion did not start without prior influence and history, even though we are led to believe otherwise. In order to establish their own ethnic identity, they denigrated the influence of the Cushites whose land and culture they claimed unto themselves. As with all immigrants in foreign lands, they absorbed much of the natives' culture, without understanding it, and, at the same time, resented it. From these poor nomadic shepherds arose three of the world's great religions—Judaism, Christianity, and Islam, each professing to be "God's chosen people." Each succeeding religion attempts to bury the former in epitaphs of "pagan" and "unholy," attempting to intercept the flow of history with a new ethnic predominance. [Unfortunately, the flow from Egyptian to Jewish to Christian to Muslim is marked by so much ethnic arrogance and bigotry that it beggars the imagination to believe that "God" had anything to with any of them! The ensuing violence makes it even harder to imagine.]

The religions of the ancient Cushites were full of their deified ancestors and their legends and the celestial zodiac. These would become the gods, mythologies, and religions of those that followed in conquest and assimilation—the Semites, the Greeks, and the Romans. Only by careful gleaning can today's students discover the "roots" of civilization, its laws, sciences, and religions, in the lives and times of the ancient Cushites, from Gen. 6:4, "the heroes old, the men of renown."

To quote Gerald Massey, "The unbelievable and fantastic in the Bible are precisely those elements that were originally mythical, then made historical. *Revealed religion is but unrevealed mythology"* (ref. 16, p. 130).

11. African Influence on Judeo-Christian Mythology*

<u>More human history occurred before the Book of Genesis than since.</u> Little does the young Afro-American realize when he proudly wears the shirt emblazoned with "Before History There Was Black History." The holy books of the Jews, Christians, Muslims, and Hindus do little justice to those who preceded. Yet, the roots of Cushite mythology and prehistory infuse the holy writings. The meanings of many of the ancient names have almost been lost, others have been changed in translations, at least in meaning if not in fact, places are now obscure, and religious history has been intertwined with ancient Egyptian mythology without explicit acknowledgment. Figure 9 is a list of a few of the ancient Hebrew names and their literal translation in Egyptian. This provides a clue about how strong was the Nile influence on the Hebrew mythologies and the Semitic language.

Ancient totemism flowed into the zodiac and astronomy adopted the symbols of nature. The stars were given names of earthly things and people and their very regular movements across the sky were given mythological proportions. Knowledge of astronomy and the ability to predict seasonal events on earth gave the priesthood enormous power. It seemed as though someone somewhere had control over events here on earth and only a handful of priests were in contact with those supernatural powers.

The priests/astronomers undoubtedly were creative in beguiling the people with stories about why things happened the way they did. Astronomical observatories turned into shrines of religious and political importance.

Every king adopted his own favorite zodiacal symbol, every age had its own zodiacal sign, and every "wannabe" king or priest bastardized the whole process for his own personal gain.

Relics of our ancient past are littered throughout our cultures, in our languages, our customs, our laws, and our religions. Judeo-Christian mythology, in particular, is heavily infested with the zodiac and the cycles of the solar, lunar, and stellar movements—the science and religion of the ancient Cushites and Egyptians millennia before the Book of Genesis.

*Material on the development of ancient religion adapted from ref. 10 and 16.

Important religious holidays, including the birth and death of Christ, coincide with specific annual solar events. For instance, the birth of Christ coincides with the Winter Solstice, the annual "rebirth" of the sun. His death and resurrection coincide with the Spring Equinox, when the sun "crosses" the stellar equator the spring plants "rise up from the dead" earth. The twelve tribes of Israel and the twelve disciples are one and the same as the twelve signs of the ancient Cushite Zodiac. The four corners of the Israeli encampment are the four cardinal points of the zodiac as seen in the names of Reuben, Judah, Ephraim, and Dan. As Gerald Massey remarked, after thirty-six years of research, "Undoubtedly, there is some very slight historic nucleus in the Hebrew narrative, but it has been so mixed with myth that it is far easier to recover the celestial allegory . . . than it is to restore the human history" (ref. 11, p. 365).

Viewed with a cool historical eye, there has been a series of steps that have taken the world from the ancient Cushite worship of ancestors and the Zodiac to the present. Moses was not a Hebrew; he sounds more like a dissident Egyptian priest bent on reform. The Ten Commandments are but a summary of The 42 Declarations of Innocence from the Egyptian Book of Maat, from about 3000 B.C., as seen in Figure 10. Christ was not a Christian; he was a Hebrew also bent on reform. His life story is identical with that of the "Egyptian Christ," Heru, also from about 3000 B.C., as seen in Figure 11, taken from the Temple of Luxor, circa 3000 B.C. just as years later, Luther would lead a reform of the Roman Catholic Church that would end up as Protestantism. Relics of the preceding religions remain intact in each step, even though we have lost track of their meaning and know not where they came from.

This should not be construed as heresy; the flow of history and the flow of religion coincide when we peer beyond the dogma imposed by the self-serving priesthood. It is a common tragedy of human thought and belief that when an institution is being reformed, we feel compelled to debunk the previous philosophies. The Hebrews made the Egyptians look evil in order to justify their exodus (or expulsion). The Christians made the Hebrews look unholy because they would not accept Christ as their savior. In the same manner, the Protestants were obligated to accuse the Catholics of grievous wrongs. It never ends, this amputation of our roots to justify change, the roots that are at the foundation of our "new" culture. Progress? Maybe, but at a high price—the loss of history and credit to those who paved the way for succeeding generations.

The age-old conflict between myth and history, between faith and

Having spent several hundred years in Egypt, the Hebrews adopted most of the Egyptian culture including the language and religion. Nothing so exemplifies this as the names of the Hebrews in the Books of Moses and their literal translation in the Egyptian language from where they came. The following is a short list of Hebrew names and their Egyptian meanings (ref.16, p143-170, citing ref.33 & 34):

Hebrew Name	Egyptian Translation
Adam	father of mankind
Eve	great mother serpent
Noah	flood irrigating the fields
Shem	traveler or nomad
Japhet	islands of the north
Ham	black
Abraham	servant of the will of Ra(sun-god)
Isaac	place of burnt offering
Israel	place that Ra created
Moses	drawn from a sea of reeds
Aaron	bull-calf
Sinai	stone of fire
David	to smite by flinging
Rachel	wise of speech

This list clearly establishes the Egyptian connection with the Hebrew stories of old. The Afro-Asiatic language as it appears in Hebrew is unmistakably African.

The Hebrew stories of these people and places, written down at least a thousand years later, reflect these Egyptian names and their meanings. The story-tellers probably lost track of where the names came from but they still knew their meanings and put those into their stories. They embellished their history with all that they had left, the meanings of the names. When and why this Egyptian connection was broken is the subject of the next chapter.

Figure 9. **Hebrew/Egyptian Names and Their Literal Translations**

1. I have not done iniquity.
2. I have not robbed with violence.
3. I have not stolen.
4. I have done no murder; I have done no harm.
5. I have not defrauded offerings.
6. I have not diminished obligations.
7. I have not plundered the Netcher [God].
8. I have not spoken lies.
9. I have not snatched away food.
10. I have not caused pain.
11. I have not committed fornication.
12. I have not caused shedding of tears.
13. I have not dealt deceitfully.
14. I have not transgressed.
15. I have not acted guilefully.
16. I have not laid waste the plowed land.
17. I have not been an eavesdropper.
18. I have not set my lips in motion (against any man).
19. I have not been angry and wrathful except for a just cause.
20. I have not defiled the wife of any man.
21. I have not defiled the wife of any man (repeated twice).
22. I have not polluted myself.
23. I have not caused terror.
24. I have not transgressed (repeated twice).
25. I have not burned with rage.
26. I have stopped my ears against the words of Right and Truth (Maat).
27. I have not worked grief.
28. I have not acted with insolence.
29. I have not stirred up strife.
30. I have not judged hastily.
31. I have not been an eavesdropper (repeated twice).
32. I have not multiplied words exceedingly.
33. I have not done neither harm nor ill.
34. I have never cursed the king.
35. I have never fouled the water.
36. I have not spoken scornfully.
37. I have never cursed the Netcher [God].
38. I have not stolen (repeated twice).
39. I have not defrauded the offerings of the Netcher [God].
40. I have not plundered the offerings to the blessed dead.
41. I have not filched the food of an infant, neither have I sinned against the Netcher [God] of my native town.
42. I have not slaughtered with evil intent the cattle of the Netcher [God].

Figure 10. **The Egyptian Admonitions of Innocence**

knowledge, never ends. Too many of us run off with the latest fads as though nothing existed before. Mythology is a dirty word in modern thought, even though we don't recognize the unspoken mythologies of our own generation and where they came from.

Many of our thoughts and values today were once those of an ancient Cushite who passed them on to an Egyptian who passed them on to a Hebrew who passed them on to a Christian and so on, just like the air we breathe and the water we drink. <u>Practically all of our laws and religions, our values and beliefs, are but old wine in new flasks.</u> The human experience has never changed. Just because we ride around in airplanes instead of on camels doesn't mean that we have advanced culturally. There is a common denominator of the human experience that never changes, call it God, call it spirit, call it genes, call it what you want; we are all intimately related in body and spirit with each other as well as to everyone who went before.

Figure 11. **The Egyptian "Christ Figure"**
The following is a copy of the scene from the Egyptian Temple of Luxor (circa 1380 BC) that depicts the story of the Egyptian "Christ figure", Heru (circa 3,000 BC) (ref.25, p95, ref.16, p188, ref.11, p757). The story of The Annunciation (1), Immaculate Conception (2), Virgin Birth (3), and the Adoration by the Three Magi (4) are depicted on an ancient temple carving. The story continues that the Egyptian "Christ figure" left home at the age of seventeen to study for the priesthood, returned at the age of thirty, was crucified, and returned from the dead. **The association of this ancient Egyptian "Christ figure" with the Christ of the historical era is impossible to ignore**.

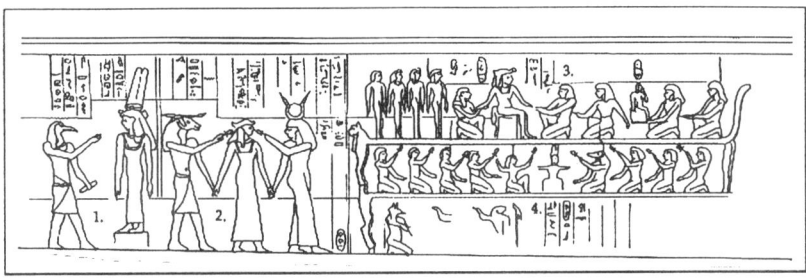

Figure 12. Earliest Caucasian Migrations
[Years B.C.]
(adapted from ref. 4, 17, & 27)

12. When and Where Did the Aryans/Semites Change the Cushite/Egyptian Mythology into Judeo-Christian Mythology and Make It Sound Like Their Own?

Somewhere, sometime, somehow in our distant past, Our Common African Genesis and the glories of the ancient Africans got "whitewashed." Let's get a historical perspective.

About 2500 B.C. the Caucasians started migrating out of their northern homelands to the south, the west, and the east, as shown in Figure 12, the author's best guess. The Greeks got to Greece around 2000 B.C. The Romans got to Italy about a thousand years later as did the modern Indians get to India (ref. 4, p. 10). The Akkadians overran Sumeria around 2371 B.C. (ref. 27, p. 85; ref. 46, p. 94). They or their kind must have kept on moving, because in 1783 B.C. the Hyksos "shepherd kings" took over northern Egypt. The Egyptians didn't kick them out until 1550 B.C. [Historical accounts about the Hyksos reign vary greatly on its duration. Estimates vary from over 100 years in ref. 46, to 150 years in the *Encyclopaedia Britannica* (ref. 31), to about 250 years from biblical sources, to 550 years from Ms. Houston's (ref. 17) references.] An Historical Timeline is shown in Figure 13 that depicts the spread of Caucasians out of their homelands.

The history of the Caucasian migrations out of the north, including the invasions of the Akkadians and Hyksos, coincides with the Books of Moses and the tales of the Hebrew patriarchs. Abram came out Chaldea about 2000 B.C. (Gen. 11:31). Joseph would not have risen in Egyptian government circles in an Egyptian regime (Gen. 41:41). Joseph rode to greet this father on a chariot (Gen. 46:28), a Hyksos introduction to Egypt (ref. 37, p. 214). The story about when "a Pharaoh came that knew not Joseph" (Exod. 1:8) coincides with the Egyptian return to power. The hated Hyksos were then run out of Egypt in a mass exodus (with numbers estimated at 200,000 or more) and are reported to have gone north into Canaan and settled around Jerusalem (ref. 16, p. 140, ref. 37, p. 129, ref. 36, p. 61). Coincidence?

It seems that the Hebrews were closely related to the invading Caucasians from the north. In fact, it seems that they were one and the same. **The**

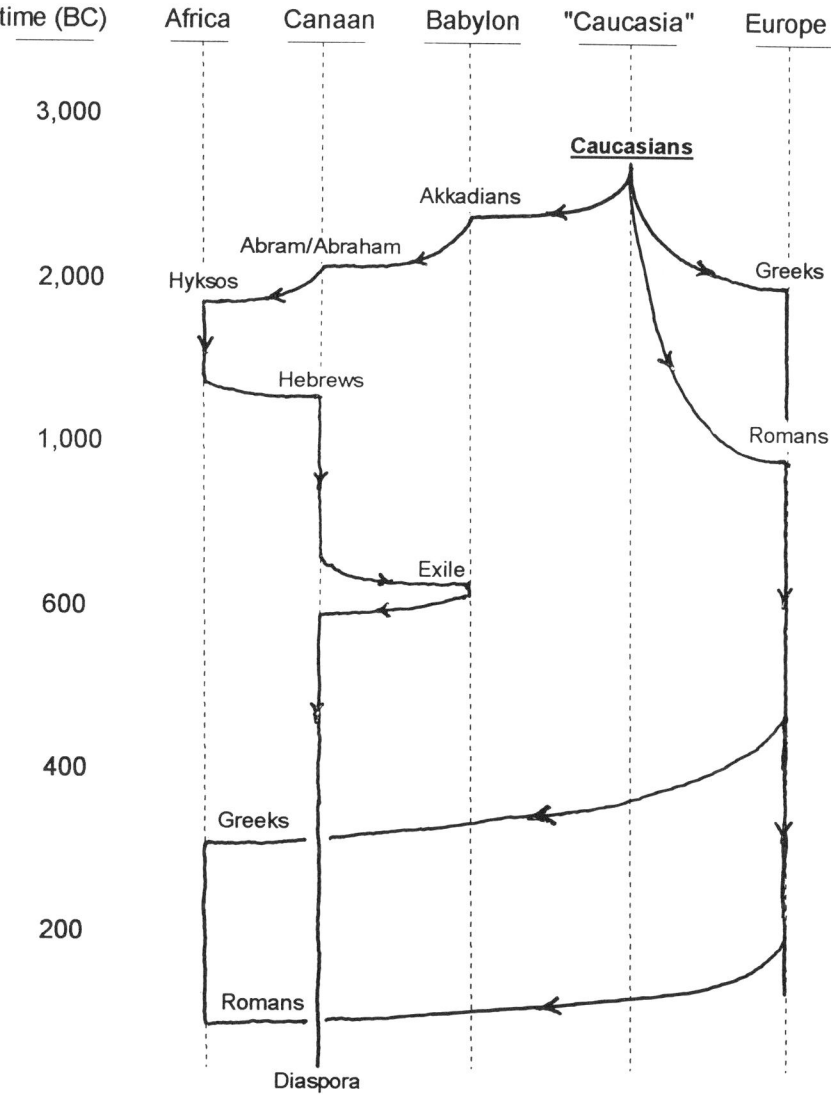

Figure 13. **Historical Timelines**
Migrations of Caucasians out of "Caucasia" (S.W. Asia)

Hebrews and Hyksos history in Egypt overlaps by 250 to 400-plus years (derived from ref. 27, p. 39 & ref. 36, p. 61). <u>The entry of the mythical Abram into Canaan about 2000 B.C. marks the first documented appearance of white men into that region and the subsequent blessing by his god seems to be an attempt to legitimize what followed.</u>

Of course, none of this was written down for about 1,400 years, long after the Hamitic population was "displaced," absorbed, or, simply, wiped out. Since none of the tales were new, we can now ask when, why, and how did the white predominance get into the African mythology? Or, to put the question another way, why was African mythology used to describe the history of the white man's invasions and takeover of the Hamitic civilizations? The two got intimately related; the question is where, when, and, in particular, why?

The flood story was ancient in the Middle East when the Caucasians showed up (ref. 7, p. 31). When and why did they add the tidbit about Noah cursing [black] Canaan to be the servant of [Caucasian] Japhet and [Semitic] Shem? (Gen. 9:25–27). It adds nothing to the flood story and why there was a flood. The Books of Moses tell only of the great civilizations of the sons of Ham and his grandson, Nimrod, the founder of Babylonia. There is no mention of such achievements by Japhet and Shem. <u>Why is this obvious Hamitic predominance at the dawn of history so overlooked when it's in the Hebrew gospels?</u> Why would invading nomadic shepherds so denigrate the founders of the civilization that they inherited or stole?

Let's look at the sojourn of Abram/Abraham in Canaan and Egypt (Gen. 12:10–20; Gen. 13:1–2; Gen. 20:1–16). This may sound like a truly tasteless joke, but this is his story according to the Holy Scriptures. First, he travels down to Egypt, tells his wife to pretend that she's his sister, pimps her off to the local officials, and they reward him with a lot of money and livestock. He goes back to Canaan a very wealthy man, tries to pimp his wife again, joins the local church of the Hamitic sun god, Ra, has the ancient Cushite ritual of circumcision performed, changes his name to Ab-Ra-Ham, gets his wife's black maid, Hagar, pregnant, and later kicks her and their son out into the wilderness to starve [history's first single black mother!]. Yet, his god finds him to be so righteous, that he gives him and his progeny (<u>except the black one</u>) all of the land from the Nile to the Euphrates.

Is this a dirty joke or not? This would be adult-rated in our time! Whoever put this story into our Holy Books had to have a twisted sense of humor or was just plain "twisted." It had to be a best-seller in Sodom and

Gomorrah! The author(s) kept stressing the idea that it was the local aboriginal Hamites who were wicked. Remember that this appearance of the first white man in Canaan was written down 1,400 years after the alleged fact. Even then, how the author(s) could have dreamed up such a story is dumbfounding to the modern mind. [My apologies to my Jewish friends; Abraham is the patriarch of the Jewish people and his holy blessing is the basis for their claim to Israel.]

The history of Moses is clouded. We have three different versions of the story to assess. The first version is tied to the exodus of the Hyksos when they were run out of Egypt (ref. 16, p. 136–40). The second version is that of a dissident Egyptian priest leading his followers out of Egypt, voluntarily or involuntarily (ref. 16, p. 140). This version was put forward by Manetho, an Egyptian priest commissioned by the Greeks in 350 B.C. to document the history of Egypt. Most aspects of his Egyptian history have been confirmed by other sources. Another variant of this story claims that Moses could have been a disciple of the Egyptian pharaoh, Akenaton, history's first acknowledged monotheist (ref. 16, p. 143; ref. 13, p. 111). This version has merit, since Akenaton's reign was at or just prior to the biblical Exodus. The third version is, of course, the biblical account (Exod.) [The author has not seen any reference to the Hyksos or a remnant of the Hyksos having been enslaved by the returning Egyptian rulers, even though the biblical account emphasizes such an episode.]

One notable feature of the biblical account is the description of the people whom the Hebrews encountered in Egypt and in Canaan; that is, the people that they describe are totally Cushite (Exod. 3:8). There is no mention of the Hyksos, their own Semitic kind. <u>That makes the exodus of Hebrews look like it was the first or the only one out of Egypt into Canaan, which makes them look like the Hyksos.</u> A later exodus, say around the time of the generally accepted biblical account, about 1290 B.C. would have resulted in the meeting of Hebrews with Hyksos, in Canaan or on the way there.

Whatever happened, it is generally agreed that a powerful Hebrew leader whom we know as Moses did lead the Hebrews/Hyksos out of Egypt and started the Hebrew tradition that has been passed down through the ages and remains the foundation of the Jewish, Christian, and Islamic religions.

However, this Judeo-Christian-Islam system of laws and mythologies is totally Egyptian in its content. The Ten Commandments are but a summary of the 42 Declarations of Innocence from the Egyptian Book of Maat, circa 3000 B.C. The other civil and hygienic laws are those of a very old, refined, civilization and have been seen in the monuments of the ancient Hamitic cultures of Egypt, Midia, Canaan, and Sumeria, for example, the Code of Hammurabi, found in Chaldea. It is not surprising that ancient historians believed that Moses was highly educated in the Egyptian priesthood. Could he have been one of the Hyksos? It seems doubtful that one of the hated Hyksos shepherd kings would have garnered this education during a military occupation. Yet, the Hyksos were in charge, politically, for 250 to 450 years and there were undoubtedly some close relationships between them and some of the Egyptians. One would think that the Egyptians would have accepted death before they would hand over their cherished religions, mythologies, and elaborate civil and medical codes to hated foreigners. History says that the Egyptian ruling class (including the priests?) fled south into Cush during the Hyksos reign. If, indeed, Moses was from the line of Shem as stated in the Bible, and he was educated as an Egyptian priest, then it's difficult to imagine how he got his education.

The Bible does tell us of another source of education (Exod. 2 & 11). Whatever Moses' ethnic origins, he was taken in by a Midianite priest and married one of his daughters. There he spent twenty-some years, no doubt tutoring under his father-in-law and his wife. Later, when he was leading the Hebrews out of Egypt, he was criticized by his sister for being married to a Midianite. However, he did ask his brother-in-law, Hobab, to lead them through the wilderness. He also asked his father-in-law for advice on how to manage his large organization and he took that advice, hence the twelve tribes organized and named after the Cushite zodiac.

The evidence supports different versions of the Moses story, even though there is a large difference between the Hebrew religious version and those of historians. The Hebrew story was obviously written by people who regarded themselves and Moses to be from the line of Shem, i.e., Semites, but the religious symbols, mythologies, and laws are Egyptian. The Semitic authors seem to have had their own "ax to grind" in order to justify their exile from Egypt and their overrunning of the Hamitic lands of Canaan.

Whether they intended to or not, some highly educated Egyptians and/or Midianites gave the Semites an education in their history, mythology, social laws, and religion. If the people whom we now call Hebrews were the invading Hyksos "shepherd kings," then during 250–450 years, they must have absorbed a lot of Egyptian culture, language, and religion. When they were eventually run out of Egypt, they must have been extremely distressed. Their political and religious leaders must have had a difficult time ridding them of the religious loyalties that they had acquired during their stay in Egypt. [Consider the feelings of present-day Americans if the American Indians staged a comeback and ran them all back out of the country. They would be people without a country, without homes, jobs, etc. Their feelings would be dominated by resentment, fear, and uncertainty.]

The Semitic authors of this phase in history gave no direct credit to the Hamites for their contributions. On the contrary, they describe them in loathsome terms, like someone who was very resentful. The story about being slaves to the Egyptians sounds like a psychological way of coping with what the Egyptians did to them. In Exodus, they do say that the Egyptians disliked them because they were lowly shepherds. <u>The close similarity between the Hebrews' descriptions of themselves as shepherds and the historical name of the Hyksos as "shepherd kings" is too much to ignore.</u> **Also, the Hebrews make no mention of the Hyksos, only of themselves, which makes the association of the two even stronger.** They should have crossed paths if they were different people. To repeat, their presence in Egypt overlapped by 250 to 400 years. Josephus, an early Jewish historian, also said that the Hebrews were descended from the Hyksos (ref. 37, p. 99).

In fact, these stories were written down so much later in time after the alleged events, that the authors and the Hebrew people, in general, may well have lost track of their real origins. In lieu of the facts, they may have simply written the stories around their own heroes, using the Egyptian language and symbols that they inherited, without knowing where any of it came from. **Abraham is a personification of the first migration of Caucasians into Canaan and Egypt.** Their reference to this first migration says that they knew nothing of their origins and who they were or what happened to them before they got to Canaan and Egypt. **There is little doubt that Caucasian/Semitic history started when they came into contact with more civilized people, the black Hamites.**

The intimate contacts with the Hamites are everywhere in the Hebrew Holy Books. Yet, they give credit only to their god for the magnificent system of beliefs and laws that they inherited and, now, embrace. Without an ancestry and culture of their own, they can hardly be blamed for this "slight" oversight. Only their God is their ancestor. There is no history or culture acknowledged before then. Mankind and mankind's "true religion" begin with the white man. They may use the language and symbols of previous cultures, but they can only acknowledge their existence in negative terms.

The man who led the Hyksos/Hebrews out of Egypt must have been highly insulted by their ignominious exile.

Let's look at the next major step in the Judeo-Christian culture that gives clues about its origins; that is, what are the African influences on Christianity? The answer is—total! The basic tenets of the Christian mythologies start with 1) the Annunciation, 2) the Immaculate Conception, 3) the Virgin Birth, and 4) the Adoration, including that of the three Magi. This story is not new as witnessed in the Egyptian Temple of Luxor, circa 3300 B.C. There, the same story is told, identically! The Egyptian "Christ" was also conceived in an identical manner and received by the three adoring Magi!? He left home at the age of seventeen, studied for the priesthood, returned home at the age of thirty, was crucified, and came back from death! So, what has the biblical account added to the story? Nothing, except that the bloodline to Mary, Christ's mother, is, also, from the line of Shem. Once again, the "line of Shem" asserts its historical and religious predominance! Egyptian mythology has been superimposed on Semitic history as though it happened to the Semites and no one else.

Who are the Semites? Their appearance coincides with the migrations of Caucasians from the north all along the areas of southeastern Europe and southwestern Asia. Only in *this* region of the Middle East are the northern invaders given a separate racial affiliation. They seem to be the mixed descendants of the invading Caucasians and local Cushites, originally a mixture of the black Sumerians and the white Akkadians, and later, a mixture of Hyksos with Egyptians and Midianites, as seen in the wives of Moses and the Twelve Tribes of Israel. They glibly identify the father of the Arabs as the bastard son of Abraham and his black Cushite maid, Hagar, but they don't dwell on their own interbreeding with the black natives of these areas. Their languages are called Afro-Asiatic or Hamitic-Semitic. However, since the Caucasians were latecomers and acquired civilization, culture, and religion from the Cushite natives of Sumeria, Canaan, and Egypt, it must follow that the languages were and still are dominated by the Hamitic tongues. Joseph Greenberg, a modern linguistic expert, believes that the mother tongue of these Afro-Asiatic languages came from Ethiopia (ref. 40, p. 728). **To ascribe a separate racial and cultural identity to the Semites is a way of removing the obvious African contributions to the people, culture, language, and religions. Even a modern-day reference with no formidable credentials, Microsoft En-**

cyclopedia *ENCARTA*, says that the Hebrew language was probably Canaanite in origin (ref. 43, Hebrew Languages).

By all accounts, the people whom we know as the Phoenicians were what was left of the original Canaanites (ref. 7, p. 151; ref. 27, p. 88), driven north by the Hebrews on the southeast, by the Philistines on the southwest, along the coast, and by the Assyrians on the east. Phoenicians are credited with giving the world the first phonetic alphabet (ref. 4, p. 31); however, by now they are labeled "Semites," so as not to associate such an accomplishment with the Hamitic black aborigines.

What we are left with as Judeo-Christian culture is a fuzzy oral "history" of the earliest white Hebrews held together with Cushite mythology from a much earlier time; where history was too fuzzy, the ancient celestial allegories were used to fill in. The names of the ancient Hebrews and their literal Egyptian translations were turned into the stories that we now know as The Books of Moses. The two histories and mythologies were then overlapped, a faint picture of the earliest white incursions from the north told with the rich mythologies of the millennia of Cushite experiences and accomplishments.

Whatever the perceived rationale that we may see for all that's happened, the myth of Ab-Ra-Ham set the stage for the Caucasian/Semitic Judeo-Christian mythology. With this mythology cloaked as a bona fide religion, the white man could take over the Holy Land and every other land populated by "colored" people with "colored" cultures and inflict their language and bastardized religion on the survivors at the point of a sword.

The exile of the southern Israelites in Babylon around 600 B.C. resulted in their bonding together as never before. Then, the Hebrew Books of Moses, as we know them, were collected and organized into a religious document (ref. 27, p. 271) with a mythology justifying the white takeover of the land but told using the symbols, zodiac, and mythologies of the ancient Cushites. They undoubtedly were heavily influenced by the Babylonians who were influenced by the previous Sumerians. This probably is when the first Caucasian/Semitic overlay of the Cushite mythology occurred. There in Babylonia, they must have obtained the genealogy of their race and heard the old flood story of that region (ref. 7, p. 31). There they must have added the vignette about Noah cursing his black grandson to become the servant of his white ones and added that God blessed Abraham with the gift of that whole area.

If the Babylonian exile resulted in the first Caucasian overlay onto the Cushite mythologies, then the next overlay occurred when the Greeks orchestrated the translation of the Hebrew texts into Greek by the Septuagint (ref. 7, p. 293). This occurred in Alexandria, Egypt, around 285–246 B.C. Then came the Latin version under the Romans and so on.

<u>Judeo-Christian mythology has drawn a curtain over the rich culture of the ancient Africans from which it drew its religions. Now, the world is expected to believe that "God" showed up in Canaan about the same time as the first white men, gave the whole territory to them, and damned the sons of Ham to servitude to them. The result has been a self-serving religion for the white race that blessed their subsequent subjugation of the "colored world."</u>

There is no nice way to debunk this historical aberration. The intent is obvious; the historical consequences—calamitous. There is no way to measure in words the subsequent suffering, murder, pillage, and destruction of the "pagan" black, brown, yellow, and red people of the world. With their "holy" blessing, the Caucasians and Semites rampaged around the world cloaked in their borrowed religions, ignorant and seemingly uncaring about where they came from, where they got their civilization, if one can call it that, and who and what they were destroying. Their inferiority is best measured in their pretentious racist attitudes, rooted in ignorance and fear.

The answer to the original question of this chapter then seems to be the following:

From the beginning of the Caucasian/Semite invasions of Cushite lands, they appropriated the Cushite culture. <u>They have nothing good to say about their Cushite teachers; **however, imitation is the best form of flattery.**</u> "White history" began when they first encountered the Black civilizations and their "history" and religions reflect that association. But the fact has to be uncovered from the overlay of the myth of white predominance in their religious and historical records and their insistence on a God-given right to believe that. When, where, and why the white invaders took over the black religions is lost in the dim records. Its first appearance is in the story of Noah, then Abraham, then Moses. But as soon as they had their own god, they felt compelled to take whatever they wanted from whomever they wanted. An ethnic blessing was all that their greed and violence needed. The priests and scribes sugar-coated the borrowed religions with the wisdom of the ages, universal platitudes, and promises of salvation in order to salve the self-righteous, but that does not cover up the ethnic intent.

What at the time was a local struggle for territory between the invading Hebrews and the native black Hamites was documented by the invaders from their point of view. The resulting documents became the foundation for three religions, allegedly inspired by God, but replete with negative portrayals of their black adversaries, thus creating a racial stereotype that has clouded all of Western culture since. When Moses denied his Egyptian culture, education, and religion, that set the tone for Western attitudes about Africans and African contributions to history, culture, and religion ever since. Vignettes about Noah and Canaan, Abraham and Hagar, Moab and Israel, Sodom and Gomorra, and Jezebel and Elijah, to name a few, only amplify the not-so-subliminal signals from our ancient past. Once these ideas were imbedded in our religions, they crept into our values and languages, thereby effecting every aspect of Western thought and actions, most often without us even knowing it.

We have been "programmed" to believe the myths of the Hebrews and ignore their excesses and exaggerations. For example, in the fall of Jericho (Joshua 6.20), "When the trumpets sounded, . . . and the walls came tumbling down," sounds like great storytelling. But when the Hebrews " . . . destroyed with the sword every living thing in it—men and women, young and old, cattle, sheep, and donkeys," the story becomes grisly, the "ethnic- cleansing" of Canaan. To have made a religion out of such tales is a tragedy yet to be acknowledged. We remember the ridiculous miracle, how the "walls came tumbling down," and forget the all-too-real slaughter of innocents.

It is probably a mistake to teach the Torah and the Bible as actual, detailed history of real people. The original storytellers spoke in parables and allegories, but later scribes and translators were too distant ("culturally-impaired") in time and language to accurately interpret the earlier "slang" and translated most of the stories as real deeds of real people. What at earlier times may have been entertaining, even humorous and off-color, spoken tales around the campsites and villages, theatrically embellished by each story-teller, became locked in writing a thousand years later. Whatever a Moses or a Joshua may or may not have done was superseded by zealous priests and scribes. This became the "word" of the one true God. We are now expected to believe that the founders of civilization, the black sons of Ham, the Hebrews' teachers, are evil and that their detractors are God's chosen.

13. Summary

Modern man originated in Africa, lived there for 100,000 years, then migrated around the world during the last ice age. Before and during the ice age, the Sahara was not a desert; it was fertile hospitable area, a veritable Garden of Eden. As the ice receded north, the Sahara became hotter and drier. This triggered migrations east, north, and south into areas more comfortable, with more food. Within the Sahara region, only the Nile River ended up with fertile enough conditions for hunters and, later, farmers. As the Sahara grew in size eastward, the first people moved east ahead of it and ended up at the head of the Nile, in what is now Ethiopia and Sudan, the biblical land of Cush, where geneticists now find the remnants of the aboriginal human beings.

The first civilizations arose in the land of Cush, came down the Nile, spread east into Arabia, Sumeria, and India, and north into Canaan. These people were black Africans who had developed the skills for civilization—agriculture, metal-work, tool-making, architecture, and, not the least, social order. They passed on the stories of their beginnings, the paradise they left behind, their "heroes of old, the men of renown," and the wonders of nature. About 4,000 to 5,000 years ago, a northern white race, nomadic shepherds, spread south and began to plunder the rich towns of the Hamitic natives. They were overpowering in their violence and greed.

The Hebrews/Hyksos were probably an offshoot of these early white invaders of the Euphrates valley. Abraham came from that region about 2000 B.C. When they came into contact with the aboriginal Hamitic people who had developed the area, the Sumerians, the Canaanites, the Midianites, and the Egyptians, they absorbed their rich and successful culture. They even tried to blend in, learned the language, joined the religion, and continued their shepherds' trade. But, they were never accepted as equals. They were considered inferior and uncivilized. Their greed and pride wouldn't allow this; they wanted it all. As their numbers grew, they (the Akkadians) overran Sumeria and (the Hyksos) staged an embarrassing bloodless coup in Egypt. The Egyptians didn't run them out for at least 250 years, after they had become virtually Egyptian, themselves, in language, religion, and culture. When forced out in an ignominious exile, they were distressed and distraught. They were not ready for the nomadic life of their ancestors. They did have one advantage, however, a highly educated and powerful leader who, in an attempt to control the distraught band of exiles, forged a new government and a new religion, building upon that of his Egyptian/Midianite mentors, one of whose families he evidently married into.

The trials and tribulations of this exiled band of Hyksos/Egyptian expatriates became epic; the new brand of exported Egyptian religion had a strong bonding effect on the people, as it developed under their tireless and imaginative leader. They became a people possessed to regain a predominance in the world. They retold the Cushite stories of creation, added their own short history in the Hamitic lands with a few Hyksos-centric vignettes, used the language and symbols of their Egyptian past, and embellished it all with a god who found favor with them, not their ex-friends, the Egyptians. They were "divinely" compelled to slaughter their new enemies, the hapless Canaanites, including women and children. But, when the tables were turned, their tragedies were blown up into epic proportions, masterpieces of moral persuasion.

To be honest, the Hebrew kingdoms never played a big role in the world or the Middle East. They were always dwarfed by their neighbors, who ran roughshod over them as they battled each other. They really needed a Messiah to save them and spoke of it often. In his absence, a 3,000-year-old Egyptian Christ-figure was resurrected as a Semitic savior. Another "new" Egyptian religion was then born in Canaan, also retold as though it happened only to the Hebrews, much like the preceding Hebrew religion. The "new religion" took off first, quite naturally, in Egypt and Ethiopia, where it was immediately recognized. It spread as a cult into the Caucasian world, suffering through years of persecution, before gaining total control of Europe. The whole of Europe then entered into the Dark Ages for 500 years, from about A.D. 500 to A.D. 1100, a period of heinous atrocities, presided over by the princes of the new church. The counterfeit Egyptian religion had long lost track of its roots after being paganized by its usurpers and subsequently released the violence of the white barbarians. The advances that were made by the Greeks with their Egyptian education were lost to Europe for those 500 years. Culture was finally resurrected in the Renaissance period, sparked by the incursion of African Islamic Moors in Spain, from A.D. 800 to A.D. 1492. The Moors had retained the ancient knowledge and were schooled in it.

Whatever offshoots of the Hebrew religion have arisen, their roots always go back to a land first populated by citizens of Hamitic descent, whose folklore and history became the mythologies and religions of their merciless invaders. The Hamitic Genesis may have been Hebraized, Christianity may have been Europeanized, but the Hamitic predominance at the beginning still shines through. <u>The sons of Ham indeed were the progenitors of mankind and civilization.</u>

<u>The existence of God is not in question nor is the fact that he takes on the appearance of the victorious.</u>

Postlude

So much evidence points to an African origin of our species and our civilization—genetics, linguistics, the Book of Genesis, pre-Egyptian history, and archaeology. Most surprising is the evidence from our Judeo-Christian-Islam religions where the foundations were laid deep in ancient Egyptian mythology, first by the giant of men, Moses, and next by the resurrected Egyptian Christ-figure, Heru, personified as Jesus. The evidence is incomplete in detail and not always consistent; however, a central theme coincides with the genetic data that takes us back in time before "verifiable history." There we learn that 1) we didn't evolve from apes or Neanderthals, 2) we appeared "suddenly" in Africa about 200,000 years ago, and 3) as we migrated around the world, separated from each other, and our genes mutated, our appearances changed, we developed different languages, and we developed different cultures.

Since then, we have built one culture on top of another, attempting to bury the former in ruins, but the Hamitic influence lives on. The Greeks acknowledged the Egyptians, the Romans acknowledged the Greeks, and the northern Europeans acknowledged them both. Yet today, in standard American references, the *Encyclopaedia Britannica* and *Webster's Lexicon,* the ancient Sumerians are said to be "non-Semitic" an all-too-typical guarded "white" admission that someone other than Caucasian or Semite contributed anything to anything.

The last parts of this essay focused on the historical and religious evidence, trying to find the common ground, often finding the differences. Archaeological evidence is hard to argue with because it doesn't carry the prejudices of the original owners. But, we are more than a "random sample" of the artifacts of our existence. There are stories passed down to us from ancient and not-so-ancient people, many with their own "ax to grind." Religious stories seem to have a single purpose, historical "accuracy" not being a priority. Wisps of reality creep in and out of these stories, while they abound with ethnic arrogance. The competition for control of men's minds has fostered a smorgasbord of religions, each one better than the others. The mental slavery of the church dwarfs the alleged bondage in Egypt. Jezebel was protecting the religious freedom of her people and for that heinous crime, our "religious history" brands her a villain.

Modern scholars don't often quote the works of the ancient Greeks,

Egyptians, and others, the hindsight of two thousand years being far superior. Older scholars were not so enlightened. Ms. Houston and several of the references tried to glean from the wisdom of the ancients, Herodotus, Stephanus, Homer, and others. The old recitations of events and the people involved with them were treated with respect, not ridicule, or worse, silence. The cold material data from archaeology has bred a philosophy of abstract theoreteical interpretation, no better or worse than the religious works so often impugned by its followers. Mythology has replaced mythology, the "truth" residing in the latest edition.

The one piece of evidence that no one seems to refute is the Table of Nations, the family of Noah, from the Book of Genesis. There, documented by no less than the Hebrews, the self-acknowledged family of Shem, is the outline of the African sources of Western civilization, religion, and social order, not to mention several, easily forgettable, places like Egypt, Jerusalem, and Babylon. There, the black sons of Ham, African in origin, are given credit for the rise of civilization, as we recognize it today. Yes, Mr. Forlong, we are all builders on old Kushite foundations.

This Afrocentric view of our origins has often been obscured by racial, religious, and political tensions. Many historical records were destroyed in the conflicts that ensued. The burnings of the great libraries at Carthage (146 B.C.), Alexandria (48 B.C.), and Serapes (A.D. 389) did untold damage to our knowledge of the ancient world, a de facto amputation of our African roots. Afterwards, while Europe languished in the Dark Ages, Timbuktu was a flourishing campus town along the trade routes of the sub-Sahara. The Dark Ages were actually alleviated by the incursion of African Islamic Moors who had retained the ancient knowledge. Columbus learned about the spherical earth and celestial navigation from the Moors. [The two go hand in hand.] European colonization broke up the last of the great African empires, after a glorious period of over 10,000 years. Yet, today, many Catholic churches in Europe still have Shrines of the Black Madonna. Ask the Pope (John Paul II) or any Polish Catholic.

Today, we need to recognize our common ties to Africa and African people. We owe the birth of our civilizations, religions, laws, sciences, engineering, and agriculture to the remote people of Cush. When we understand this Afrocentric nature of our origins and culture, our world view will be more complete and tolerant. We can then consider ourselves to be part of one race, the human race.

Author's Apologia

Let me explain myself; over forty years ago, as a very young adolescent, I walked out of the Christian church with a finality and rationale that makes me marvel at the wisdom of youth. I've never looked back. This essay didn't start out as a diatribe against the Judeo-Christian tradition, but evidently it was leading me to the very thing that repulsed me as a youngster.

I knew that the Christianity that I was raised in was sick, but I didn't know why. I knew that the "white" society that I was born into was sick, but I didn't know why. I wandered in the wilderness for forty years until I met and communed with the sons and daughters of Ham—black people. Then and there I was exposed to a broader and deeper culture, one that was very old, one that went back to the roots of humanity and civilization. I then started to unravel the false myth of white historical predominance. In case you haven't caught on by now, I'm white. <u>Yes, before "his" story, there was black history!</u>

This essay was put together piecemeal, a step at a time, over several years. I've left it pretty much as it went together, as a journey back in time, through genes, languages, history, and religion, searching for the "roots" of our Western culture. Throughout the whole process, I came to believe that there were no white people in the Old Testament. When I got to the chapter on the influence of the African mythologies on Judeo-Christian mythology, I hesitated. For some reason, I felt that I wasn't finished. I sensed that I was missing a connection. I looked again at the history of the surge of Caucasians out of their homelands about 2500 B.C. <u>I then picked up the Bible once more and, for the first time, saw through the overlay of false Caucasian/Semitic myths.</u> It had been there all of the time, but it was hard to break through a lifetime of indoctrination.

<u>I came to realize that the entire Judeo-Christian holy books were written by white men bent on promoting their own kind.</u> Their denigration of our black forefathers is very transparent once you can tell myth from mythology. The almost total use of African legends, mythology, and religions is the first sign that the contributions of the ancient Africans are immense. It also is a sign that white men had no history, religion, or culture of their own until they encountered Hamitic civilizations. There is, however, this diabolical twist to the biblical stories, that is the insertion of white predominance, a "God-given" white predominance, at that. They reset the

clock of creation to their own beginnings, created God in their own image, and even look for Noah's Ark near their own homelands. They have created a separate race of men, the Semites, in order to distance themselves from the Hamitic origins of their stolen culture.

The Hebrews developed a separate race, religion, and culture with only the help of their newfound god. The Greeks developed "intellect" all by themselves, with no inputs from anyone, thanks to their superior "white" brainpower.

However, the thieves have been caught, red-handed. A good part of their mythology, laws, and religions were lifted out of their African predecessors' civilization. We are the recipients of this stolen treasure—African mythology passed on as white history, white history passed on as religion.

The Hebrew Books of Moses mark the entrance of white men onto the world's stage. The Bible and the Koran are their license to take over.

When this revelation hit me, I had very mixed feelings. On one hand, I was elated to have these lifelong questions answered, but, on the other hand, I was extremely depressed to discover that the history of our sick, racially divided society goes all of the way back to the roots of our society, our cherished Judeo-Christian tradition.

What started out as a personal odyssey into our ancient past turned into a chilling revelation about our basic tenets, our religions. I discovered that "God" didn't show up in Canaan until the first white men showed up!? And he's on their side!?

However, we must be grateful to the Hyksos/Hebrews for preserving the wisdom of the ages, even though we must clarify its origins and peer through the veil of Semitic predominance at the dawn of history. **They have paid un ungodly price for bringing this ancient blessing to the modern world. Their tradition was born out of the humiliation of their first exile from Egypt, was bonded in their next exile in Babylon, survived their final Diaspora from Canaan, and refused to perish in the ovens of their European Holocaust. The Hyksos "shepherd kings" are still alive and well, thank you, back home in Canaan. Shalom.**

Quite frankly, I don't see any way to correct something so old and so revered. **It does seem that the world's great religions are designed around specific ethnic groups.** Then, the religions are used to enforce ethnic bigotry and sanctify a host of sins. Those sins, when inflicted on us by someone else, are monstrous, yet when inflicted by us on them, are

"God's will." I don't see religion as one of man's higher callings; rather, I see it as ethnic and racial politics, a way to get away with murder, ethnic cleansing, the Inquisition, and the Holocausts of the Maoris, the American Indians, the African slaves, and the Jews of Western Europe.

<u>Perhaps that is just the Judeo-Christian legacy—the unbridled rampage of white men across the earth.</u>

* * *

My apologies; I really never knew I was heading towards this cold conclusion. You don't have to agree or disagree; just read and discuss this topic some more. It is only in the calm rational realm of the community at large that we can find our common spiritual relationships. Racial and religious intolerance is the enemy of mankind and the germ of our destruction.

* * *

References

1. "Gene Data Place Home of 'Eve' in Africa." B. Bower, *Science News,* September 28, 1991.
2. "Genes, Peoples, and Languages," Luigi Luca Carvalli-Sforza, *Scientific American,* November 1991.
3. "Recent African Genesis of Humans, The," Alan C. Wilson & Rebecca L. Cann, *Scientific American,* April, 1992.
4. *Languages of the World, The,* Kenneth Katzner, Routledge, London & New York, 1977, rev. 1986.
5. "World Linguistic Diversity," Colin Renfrew, *Scientific American,* January 1984.
6. *In the Age of Mankind: A Smithsonian Book of Human Evolution,* Roger Lewin, Smithsonian Books, Washington, D.C., 1988.
7. *ABCs of the Bible,* Kaari Ward, editor, The Readers Digest Association, Pleasantville, NY/Montreal, 1991.
8. *Golden Ages of Africa, The,* John G. Jackson, American Atheist Press, Austin, TX, 1987.
9. *Herodotus, The Histories,* translated by Aubrey de Selincourt, revised by John Marincola, Penguin Books USA, New York, 1996.
10. *Man, God, and Civilization,* John G. Jackson, A Citadel press Book, Carol Publishing Group, New York, 1993.
11. *Ancient Egypt: The Light of the World,* Gerald Massey, T. Fisher Unwin, London, 1907, Samuel Weisser, Inc., 1970.
12. *Tropical Dependency, A: An Outline of the Ancient History of the Western Sudan with an Account of the Settlement of Northern Nigeria,* Lady Lugard, James Nisbet & Co., 1906, Frank Cass & Co., 1964, Barnes & Noble, New York, 1965, Black Classic Press, 1995.
13. *Introduction to African Civilizations,* John G. Jackson, A Citadel Book, Carol Publishing Group, New York, 1993.
14. *From Babylon to Timbuktu,* Rudolph R. Windsor, Windsor's Golden Series, Atlanta, 1988.
15. *Ethiopia and the Origin of Civilization,* John G. Jackson, Black Classic Press, Baltimore, 1985.
16. *Echoes of the Old Darkland,* Charles S. Finch III, M.D., Khenti Inc., Decatur, GA, 1994.
17. *Wonderful Ethiopians of the Ancient Cushite Empire,* Drusilla Dunjee Houston, Black Classic Press, Baltimore, 1985 (original copyright 1926 by Ms. Houston).

18. *Origin of Nations, The,* George Rawlinson, Scribner, Welford & Armstrong, New York, 1878.
19. *Rivers of Life,* J. G. R. Forlong, Bernard Quaritch, London, 1883.
20. *History of Ancient Civilization,* C. Seignobos, T. Fisher Unwin, London, 1907.
21. Y Chromosome and the Origin of US All Men, The," Svanta Paabo, *Science,* Vol. 268, p. 1141, 26 May 1995.
22. "Absence of Polymorphism at the ZFY Locus on the Human Y Chromosome," Robert L. Dorit, Hiroshi Akashi, Walter Gilbert, *Science,* Vol. 268, p. 1183, 26 May 1995.
23. *Essentials of Earth History,* Fourth Edition, W. Lee Stokes, Prentice-Hall, Inc, Englewood Cliffs, NJ, 1982.
24. "Africa's Ancient Cultural Roots," Bruce Bower, *Science News,* Vol. 148, p. 378, December 2, 1995.
25. *Nile Valley Contributions to Civilization: Exploding the Myths*— vol. 1, Anthony T. Browder, The Institute of Karmic Guidance, Washington, D.C., 1992.
26. *Outline of History, An,* H. G. Wells, Triangle Books, New York, 1940–41.
27. *New Bible Atlas, A,* edited by J. J. Brimson, J. P. Kane, J. H. Paterson, D. R. W. Wood, Inter-Varsity Press, Leicester, Lion Publishing, Oxford, 1985.
28. *Historical World Atlas,* vol. 2, Hammond Inc., Maplewood, N.J., 1961.
29. *Canaanite Myth and Hebrew Epic: Essays in the History of the Religion of Israel,* Frank Moore Cross, Harvard University Press, Cambridge, Mass. and London, 1973.
30. "Mitochondrial DNA and Human Evolution," Rebecca L. Cann, Mark Stoneking, & Allan Wilson, *Nature,* Vol. 325, January 1987.
31. *Encyclopaedia Britannica, The,* vol. 11, Micropedia, 15th edition, Robert P. Gwinn (Board of Editors), Encyclopaedia Britannica, Inc., 1991.
32. *Webster's Lexicon.*
33. *Interpreter's Dictionary of the Bible,* G. A. Buttrick, (editor), Abingdon Press, Nashville, 1984.
34. *Egyptian Hieroglyphic Dictionary, An,* E. A. W. Budge, Dover Publications, New York, 1921, 1978.
35. *Ancient History,* George Rawlinson, Barnes & Noble Books, New York, 1993.
36. *The Holy Bible,* International Bible Society, Colorado Springs, CO, 1973, 1978, 1984.
37. *Egypt, Canaan, and Israel in Ancient Times,* Donald B. Redford, Princeton University Press, 1992.
38. *Origin of Species and the Descent of Man The,* Charles Darwin, The Modern Library, New York.

39. *Jewish Antiquities,* Flavius Josephus, Loeb Classical Library, Cambridge, Mass., 1988.
40. *Historical Researches: African Nations,* Arnold Hermann Ludwig Heeren. Cited by John D. Baldwin in "Pre-Historic Nations," Harper and Bros., New York, 1869.
41. "Ancient Egyptian Outpost Found in Israel," *Science News,* Vol. 150, October 5, 1996.
42. *Peoples and Cultures of Africa,* J. H. Greenberg, Natural History Press, Garden City, 1973.
43. *Microsoft* Encarta 97*Encyclopedia,* Microsoft Corporation, 1993–1996.
44. "Molecular Handle on the Neanderthals, A, Ryk Ward & Chris Stringer, *Nature,* Vol. 388, July 17, 1997.
45. "Distant Cousins," Roger Lewin, *New Scientist,* July 19, 1997.
46. *Timelines of the Ancient World,* Chris Scarre, editor, Smithsonian Institution, Dorling Kindersley, New York, 1993.

Index

Aaron, 34
Abraham, 12, 15, 19, 30, 34, 39, 40, 43, 44, 45, 46, 47
Adam, 12, 13, 19, 34
Afro-Asiatic, 6, 8, 34, 44, *see Hamitic-Semitic*
age of mankind, 16, 18
Akenoton, 41
Akkadians, 38, 39, 44, 48
Alexander the Great, 21
amen, 23
Amman, 21
Amun, 27
Andromeda, 21
apes, 1, 9, 10, 11, 51
Aryan(s), 7, 22, 24, *see Caucasian, European, Indo-European, white*
Asians, 1, 5, 6

Babylon, 14, 15, 19, 22, 39, 45, 52
Ba'l, 30
Bible, 31, 42, 53
Biblical Exodus, 41, *see exodus*
black, 22, 32, 40, 44, 45, 46, 47, 48, 52, 53, *see Cushitic, Hamitic, Nilotic*
Black Madonna, 52
Books of Moses, 18, 34, 38, 45, 54

Canaan, 15, 21, 24, 26, 27, 28, 30, 38, 40, 41, 43, 44, 47, 49, 54
cardinal points, 33
Cassiopeia, 21
Caucasian(s), 1, 5, 24, 28, 37, 38, 39, 40, 43, 44, 45, 46, *see Aryan, European, Indo-European, white*
Cepheus, 21
Chaldea, 15, 42
Christ, 24, 28, 33, 36, 44, 49
Christmas, 31
circumcision, 40
Code of Hammurabi, 42

Colchis, 21
Columbus, 52
Creation, 19, 30, 49
cult, 49
Cush, 14, 15, 21, 42, 48, 52
Cushite, 21, 22, 23, 24, 30, 32, 33, 36, 38, 40, 41, 42, 44, 45, 46, 49, *see black, Hamite(s), Kushite*
Cushite Zodiac, 28, 29, 33, *see zodiac*

Dark Ages, 49, 52
Darwin, 7, 9
David, 34
Declarations of Innocence, 33, 35, 41
deluge, 19
DNA, 9

Easter, 31
Egyptian "Christ," 33, 36, 44, 49, 51
Egyptian Book of Maat, 33, 35, 41, *see Declarations of Innocence*
Ethiopia, 7, 14, 15, 21, 23, 44, 48
Euphrates, 22, 25
European, 1, 3, 6, 21, 28, 52, 54
Eve, 10, 13, 19, 34
evolution, 1, 2, 9, 10, 11, 12
exile, 44, 45, 48, 49, 54
exodus, 27, 30, 38, 41, 43, *see Biblical Exodus*

family tree, 1, 3, 4, 6, 7, 10, 11, 12, 14, 19, *and front cover*
flood(s), 12, 15, 18, 19, 28, 34, 40
flood story, 12, 15, 18, 19, 40, 45
fossils, 9, 11, 12

Garden of Eden, 48
genes, 1, 6, 8, 36, 51
Genesis, 12, 15, 18, 19, 30, 32, 38, 50, 52
genetic clock(s), 1, 9, 11, 13, 18

giants of the earth, 11, 19
God, 13, 18, 31, 35, 36, 43, 46, 47, 53, 54, 55

Hagar, 40, 44, 47
Ham, 14, 15, 27, 34, 40, 45, 50, 52, 53
Hamite(s), 22, 41, 43, 47, *see Cushite(s), black*
Hamitic, 12, 13, 24, 28, 40, 43, 53, *see Cushitic, Kushitic*
Hamitic-Semitic, 6, 7, 8, 27, 34, 44, *see Afro-Asiatic*
Herodotus, 21, 24, 25, 52
"heroes of old, men of renown," 21, 22, 48
Hobab, 42
holy books, 32, 53, *see Bible, Books of Moses, Old Testament*
Homer, 36, 52
hominid(s), 10, 11
Hyksos, 26, 27, 34, 40, 42, 43, 44, 48, 54, *see Shepherd Kings*

ice age, 3, 10, 12, 15, 16
Indo-European(s), 5, 6, 22, 24, *see Aryan, European, Caucasians, white*
Isaac, 34
Israel, 7, 15, 16, 24, 27, 34, 47

Japhet, 14, 15, 34, 40
Jerusalem, 15, 38, 52
Joseph, 38
Josephus, 26, 43
Joshua, 47
Judeo-Christian, 32, 38, 44, 45, 46, 51, 53, 54, 55

King Tut, 23
Kushite(s), 22, *see Cushite, Hamite*
Kushitic, 22, *see black, Hamitic, Nilotic*

languages, 3, 5, 7, 8, 22, 34, 44, 45, 47

Manetho, 41
Mesopotamia, 25
Messiah, 49
Middle East, 5, 16, 19, 44, 49
Midianite, 32, 42, 44, 48
migration(s), 2, 3, 4, 5, 7, 15, 37, 38, 39, 44, 48
monotheist, 41
Moors, 49, 52
Moses, 8, 18, 27, 30, 33, 41, 42, 45, 46, 51, 54
mother tongue, 7, 44
mutations, 1

Neanderthals, 1, 9, 10, 18, 51
Neolithic, 8, 25
Nile, 8, 21, 22, 25, 28, 48
Nilotic, 27, 28, *see Hamitic, Cushitic*
Nimrod, 14, 15, 20, 40
Noah, 12, 14, 15, 34, 45, 46, 47
Noah's Ark, 10, 12, 15, 20

Old Testament, 53

Palestine, 26
patriarch, 19, 41
Phoenician(s), 21, 23, 24, 45
priest(s), 30, 32, 42, 46
primate, 9, 10, 16
prophets, 27
Ptolemy, 21

Ra, 34, 40, 45

Samaritans, 24
Second Temple, 27
Semites, 15, 22, 23, 28, 44, 46
Septuagint, 69
Shem, 14, 15, 28, 34, 40, 44, 52
Shepherd King(s), 23, 38, 42, 43, 54, *see Hyksos*
shepherds, 15, 22, 23, 30, 31, 48
Sinai, 30, 31, 34
Spring Equinox, 33
stone age, 16, 25

Sumeria, 38, 42, 44, 48

Temple of Luxor, 33, 36, 44
Ten Commandments, 23, 44
Theory of Evolution, 2, 9, 10, 11
Timbuktu, 52
Tower of Babel, 19
twelve tribes of Israel, 30, 33, 44

Virgin Birth, 36, 44

white, 5, 7, 9, 23, 40, 43, 44, 45, 51, 53,
   see Aryan, Caucasian, European,
   Indo-European
Winter Solstice, 31, 33

Yahweh, 27

zodiac, 28, 29, 31, 32, 33, 42, 66

Vita

Larry West; Physicist, Engineer, Teacher, And Writer
Born: March 22, 1940, Decatur, Illinois, USA
Childhood: Only child, son of farmhand
Education: Physics
B.S., 1963, University of Illinois
M.S., 1969, University of Maryland
Vocations: Engineering physics, freelance writer, musician, laborer, part-time landscaper (token WASP minority)
Last Known Residence: Atlanta and Houston (in the 'hood)
Marital Status: 3rd time, . . . 'til I get it right
Religious Preference: Unitarian
Favorite Books: *The Hebrew Books of Moses, The Wonderful Ethiopians of the Ancient Cushite Empire*
Favorite Author: Camus
Favorite Quote: $e^{i\pi} = -1$
Favorite Movies: *Who's Afraid of Virginia Woolf, The Color Purple*
Favorite Music: Ray Charles, Paul Simon, and Beethoven
Favorite Places: New Mexico and Maine Coast
Favorite Food: Any ethnic food except my own
Favorite Drink: Real iced tea, sweetened with honey and flavored with lemon and spearmint
Hobbies: Gourmet health food ethnic cooking, reading and writing (non-fiction), scientific research (fiction, bordering on nonfiction), fitness and sports, gardening

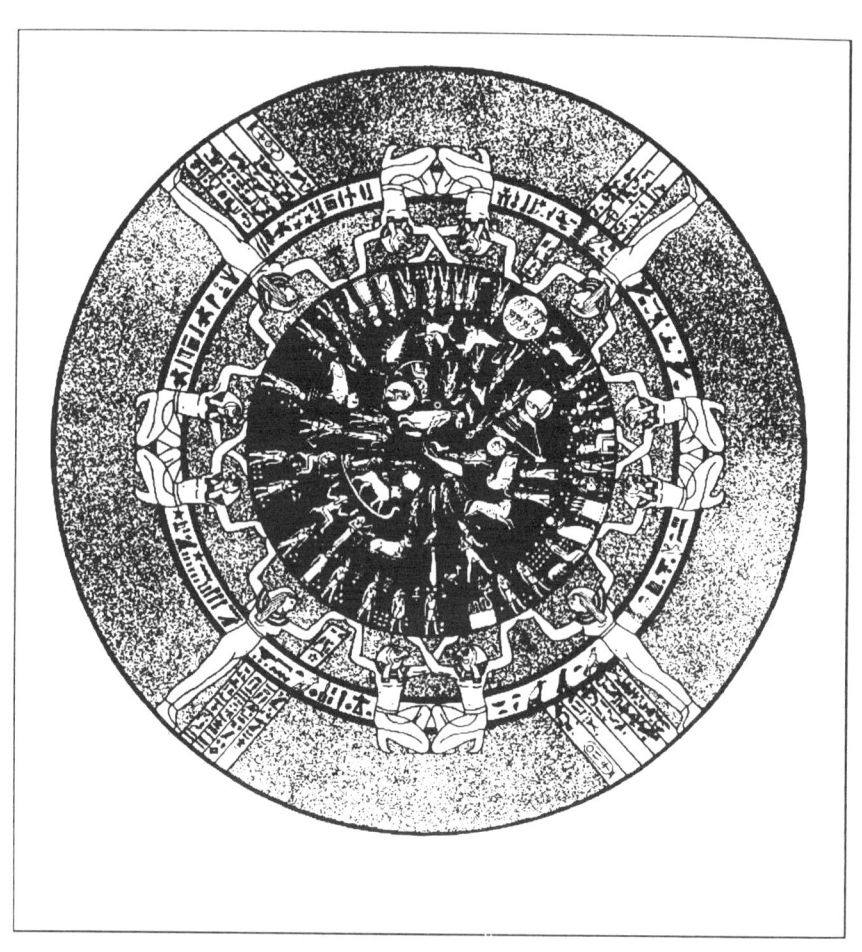

Figure 14. **Zodiac of Dendara**
(from ref. 79 and 81)